The Latin American Studies Book Series

The Latin American Studies Book Series promotes quality scientific research focusing on Latin American countries. The series accepts disciplinary and interdisciplinary titles related to geographical, environmental, cultural, economic, political, urban and health-related research dedicated to Latin America. The series publishes comprehensive monographs, edited volumes and textbooks refereed by a region or country expert specialized in Latin American studies.

The series aims to raise the profile of Latin American studies, showcasing important works developed focusing on the region. It is aimed at researchers, students, and everyone interested in Latin American topics.

Submit a proposal: Proposals for the series will be considered by the Series Advisory Board. A book proposal form can be obtained from the Publisher, Andrea Brody-Barre (andrea.brody-barre@springer.com).

Gerald Augusto Corzo Perez · Laszlo G. Hayde ·
Rabindranarth Romero-López ·
Arturo Jiménez Palacios
Editors

Aqua Vitae

Unraveling the Story of Water in Latin America

Springer

Editors
Gerald Augusto Corzo Perez
IHE Delft Institute for Water Education
Delft, The Netherlands

Rabindranarth Romero-López
Facultad de Ingenieria Civil
Universidad Veracruzana
Xalapa, Mexico

Laszlo G. Hayde
Department of Land and Water
Management
IHE Delft Institute for Water Education
Delft, The Netherlands

Arturo Jiménez Palacios
Facultad de Economía
Universidad Veracruzana
Xalapa, Mexico

ISSN 2366-3421 ISSN 2366-343X (electronic)
The Latin American Studies Book Series
ISBN 978-3-031-57733-8 ISBN 978-3-031-57731-4 (eBook)
https://doi.org/10.1007/978-3-031-57731-4

This Springer imprint is published by the registered company Springer Nature Switzerland AG
The registered company address is: Gewerbestrasse 11, 6330 Cham, Switzerland

Prologue

Latin America's landscapes and futures have been deeply affected by water in all its forms, from the icy torrents that run down the Andes to the slow flows of the Amazon River. In *Aqua Vitae: Unraveling the Story of Water in Latin America*, we investigate the complicated relationship between water and the people who live in this multicultural area. Their ability to stay alive and grow is closely linked to this resource.

Part I: *Ancient Foundations.* We begin with the ancient civilizations, where reverence for and mastery over water laid the foundations of society. From the engineering marvels of the Inca, who sculpted the rugged Andean environment to suit their agricultural needs, to the Aztecs' strategic manipulation of the lakes around Tenochtitlan, this chapter reveals how water was central to cultural identity, spiritual life, and survival.

Part II: *Visions and Viewpoints.* The narrative flows into the colonial and postcolonial eras, exploring how shifts in governance and technology brought new perspectives and policies toward water management. This chapter examines the transformation in water usage and control through the lenses of both indigenous traditions and European influences, highlighting conflicts and convergences that have shaped modern water policies.

Part III: *Modern Currents and Future Streams.* As we navigate toward the present, the focus shifts to contemporary challenges and innovations in water management across Latin America. From Brazil's efforts in the Paraná River Basin to Cuba's adaptations to climate change, this chapter discusses the ongoing struggles and strategies in ensuring water security and sustainability amid environmental and socio-economic pressures.

In this journey through time and water, "Aqua Vitae" not only charts the historical courses that have defined Latin America but also reflects on the lessons that can guide future endeavors in water stewardship. As you turn these pages, may you gain insights into the power of water as both a creator and a sustainer of life and civilization.

Contents

Part III Modern Currents and Future Streams

Part I
Ancient Foundations

Chapter 1
Water and the Aztec Empire

Mariana Castaneda-Gonzalez

Abstract The Aztec or Mexica civilization, a well-known Mesoamerican culture located in what is now central and southern Mexico, had a strong relationship with water in various aspects of their lives. Their capital Mexico-Tenochtitlan was founded over a complex watershed consisting of five interconnected lakes. This city and its innovative hydraulic structures and agriculture techniques are some of the characteristics that exemplify the Aztecs close relation with water. As for any civilization, water distribution was an essential need, the Aztecs constructed a network of aqueducts to fulfill the needs of the population. These hydraulic structures not only allowed for daily distribution of freshwater but also served as a water reserve. The Chapultepec aqueduct also known as the "great aqueduct" is a great example of such aqueducts that can still be seen in Mexico City nowadays. Yet, being a city surrounded by water also brought challenges, such as the need for structures for flood control. These led to developing a system of roads and a dike of 16 km long and 20 m wide designed to control the water levels. Another fundamental part of their development was certainly their innovative agricultural hydroponic system of chinampas that allowed providing food for more than 100,000 habitants of the Aztec capital. This article will thus describe the undeniable close relationship between water technology and development for the Aztec society, from its particular location, complex hydraulic infrastructure to its innovative farming techniques.

Keywords Mexica civilization · Hydrology · Hydraulic structures · Hydroponic systems · Chinampas

1.1 Introduction

The Aztec empire, also known as Mexica empire, is a Mesoamerican civilization that had a strong relationship with water in many aspects of their culture and day-to-day lives. This strong water-society nexus was particularly portrayed in its capital, the

M. Castaneda-Gonzalez (✉)
Ecole de Technologie Superieure (ETS), Montreal, QC, Canada
e-mail: mariana.castaneda-gonzalez.1@ens.etsmtl.ca

G. A. Corzo Perez et al. (eds.), *Aqua Vitae*, The Latin American Studies Book Series,
https://doi.org/10.1007/978-3-031-57731-4_1

city of *Mexico-Tenochtitlan*. The city of Mexico-Tenochtitlan, located in what is now central Mexico (see Fig. 1.1), was founded over an island in the lacustrine basin of the Valley of Mexico.

By the Spaniards arrival to Mexico-Tenochtitlan in the 1500s, this island covered an approximate area of about 10 km^2 with a population of about 100,000 to 150,000 people, making it one of the largest cities of the world (Becerril and Jiménez 2007; Mundy 2021). Additionally, the different towns located over the lakeshore made a greater combined population that might have reached up to half a million inhabitants (Mundy 2021).

According to Aztecs mythology, their ancestors migrated south from a mythical place called Aztlan. This migration started when the God Huitzilopochtli communicated to the people to find a new place to settle with particular characteristics. Huitzilopochtli told the Aztec people to settle and build a city where they saw an eagle standing over a cactus and eating a snake. This long migration looking for this particular place was represented in the *Tira de la Peregrinacion*, also known as Codex Boturini, presented in Fig. 1.2. The first page of this manuscript presented in Fig. 1.2 shows the mythical island of Aztlan on the left and a man leaving in a canoe (Mundy 2021). On the right, it is shown that the man travels toward a hill where Huitzilopochtli, wearing a hummingbird feather headdress, is awaiting.

The Aztecs founded this unique city and developed innovative technology to create one of the greatest empires in about 200 years after settling in the Valley of Mexico in 1325 AD. Based on old Aztec codices (as the one presented in Fig. 1.2), as well as on texts written by historians and ruins of old infrastructure that can still be observed today, the innovative water technology of this city has been studied and recognized around the world. The complex structure of the basin and location of the

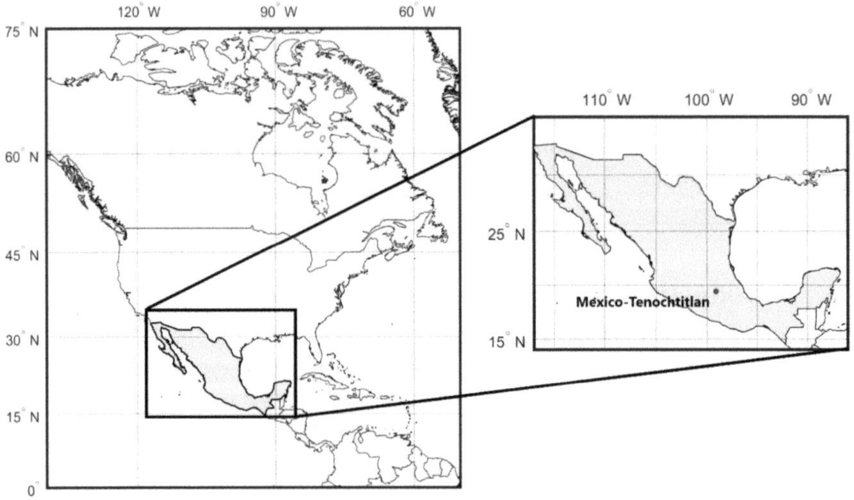

Fig. 1.1 Location of Mexico-Tenochtitlan (Coordinates not provided)

Fig. 1.2 Departure from the mythic city Aztlan. Representation painted on amate paper from the Tira de la Peregrinacion, also known as Codex Boturini, from unknown author

city brought different advantages, such as a freshwater source for the inhabitants and their crops. However, it also engendered complex issues such as managing water-level fluctuations and the impacts of extreme events such as floods (Musset 1989). The occurrence of these extreme hydrological events raised the Aztecs need to develop hydraulic infrastructure to manage and mitigate the impacts of these events.

Water was an essential and strategic part of the Aztec's progress and development, particularly for the capital *Mexico-Tenochtitlan*, their innovative water management strategies and the challenges that derived from those technologies might serve as an example to better understand and assess how water has shaped and continues to shape our societies around the world.

1.2 The Structure of the Basin of the Valley of Mexico

The basin of the Valley of Mexico is a lacustrine and endorheic catchment located in the Mesoamerican region at an altitude of about 2000 m.a.s.l. (Biar 2020). In other words, this region was a naturally closed basin with no outlet to the sea or river to drain the water, a characteristic that favored the formation of shallow lakes, lagoons and marshes. Furthermore, the impermeability of this basin's volcanic soil along with the surrounding chain of high mountains, such as the *Popocatépetl* and *Iztaccíhuatl* at 5452 and 5230 m height, favored the emergence of a system of lakes (Biar 2020; López Luján 2002). A representation of the basin is displayed in Fig. 1.3.

As observed in Fig. 1.3, this complex basin was composed of five shallow interconnected lakes: (1) Lake Zumpango, (2) Lake Xaltocan, (3) Lake Texcoco, (4)

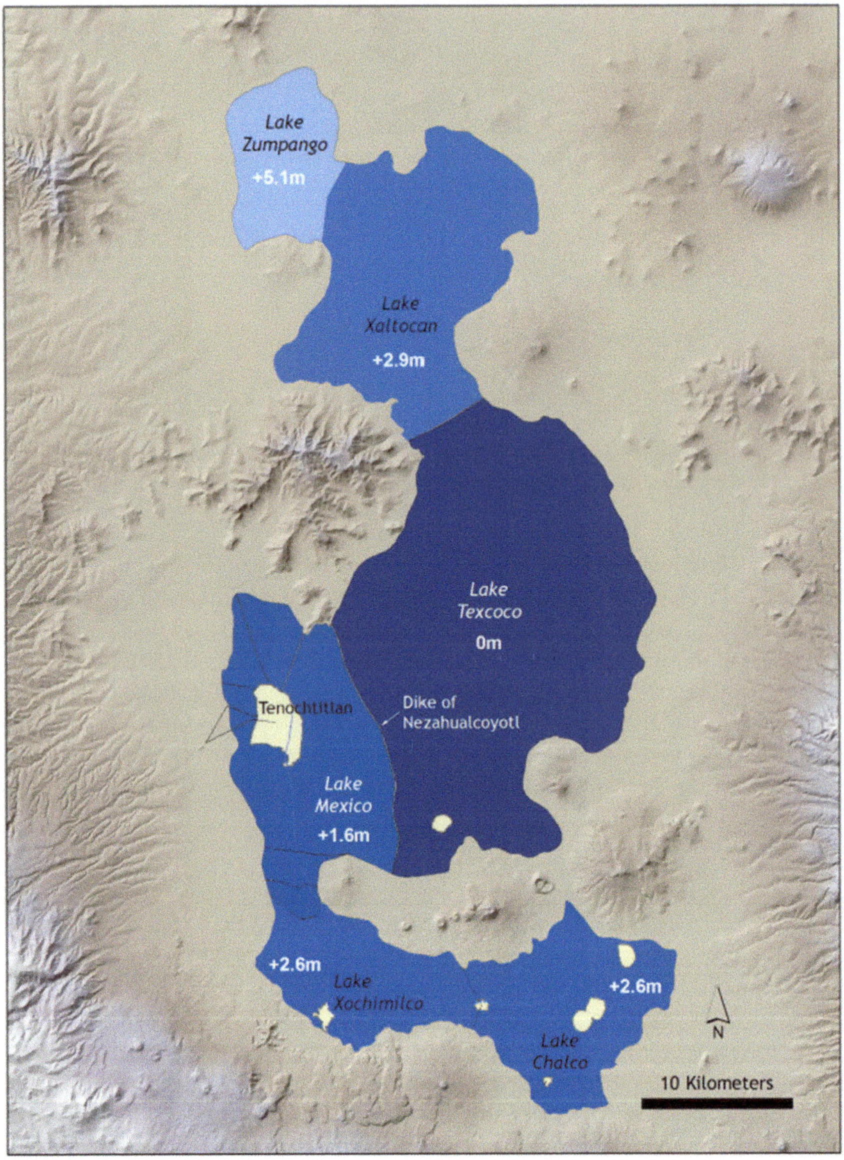

Fig. 1.3 Valley of Mexico with and lake levels with the deepest Lake Texcoco as reference. Reprinted from Luna Golya (2014), after Raynal-Villasenor (1987) (Coordinates not provided)

Lake Xochimilco, and (5) Lake Chalco. It has been estimated that the system of lakes, marshes and lagoons covered about 8000–1000 km^2 of the basin surface by the 1500s (Becerril and Jiménez 2007; Biar 2020). These interconnected lakes were also located at different altitudes, with the main Lake Texcoco located about 3–6 m below the other lakes. During the rainy seasons, the water of the basin was thus flowing from the higher lakes, namely Lakes Chalco and Xochimilco in the south and Lakes Xaltocan and Zumpango in the north, toward the lowest-lying Lake Texcoco (Revueltas Valle 2017). However, estimating the total lakes surface has been a complex task due to the water-level fluctuations of the lake.

The basin of the Valley of Mexico was a very diverse ecosystem not only characterized by endemic flora and fauna, but also by other particular hydrological characteristics such as saline water in Lake Texcoco. This lake, being the lowest, received all the water flowing from the mountains that had high concentrations of salts and minerals (Sosa-Rodriguez 2010). As all these high concentrations of salt and minerals arrived at Lake Texcoco, its water was even saltier than seawater (Sosa-Rodriguez 2010). A characteristic that was underlined by the Spaniards, as described in letter from Hernan *Cortes "This Province is in the form of a circle, surrounded on all sides by lofty and rugged mountains'; its level surface comprises an area of about seventy leagues in circumference, including two lakes, that overspread nearly the whole valley, being navigated by boats more than fifty leagues round. One of these lakes contains fresh, and the other, which is the larger of the two, salt water.*"

In terms of climate, the Valley of Mexico is characterized by a single rainy season from June to September with an irregular geographic distribution (Becerril and Jiménez 2007). During this season, the basin accumulated excessive amounts of precipitation in its aquifer that consequently generated springs around the lakes. Thus, when the dry season occurred, the springs disappeared and the lakes levels decreased with the phreatic levels (Legorreta Gutiérrez 2006). This region's climate is warm and temperate, meaning that the basin is characterized by average temperatures of about 18 °C and mean annual precipitation of 640 mm.

1.3 Mexico-Tenochtitlan

As previously mentioned, this city was located over an island in Lake Texcoco surrounded by different towns that covered about 1000 hectares (Becerril and Jiménez 2007). This city, founded in AD 1325, was composed of a geometric network of streets with houses and religious centers. This can be observed in the Map of Nuremberg (see Fig. 1.4), a representation of Mexico-Tenochtitlan ordered by Hernan Cortes.

As observed in Fig. 1.4, all houses, buildings, and "floating" agricultural lands were located around two major elements of the city, the *Templo Mayor* (i.e., the Great Temple) and the *Tenochtitlan seat* (Becerril and Jiménez 2007; Mundy 2021) where the main religious, economic, and political elements of the Aztecs were centralized. To connect the center of the city with the suburban areas, roads and bridges were built dividing the city into four main sections, (1) Atzacualco, (2) Cuepopan, (3)

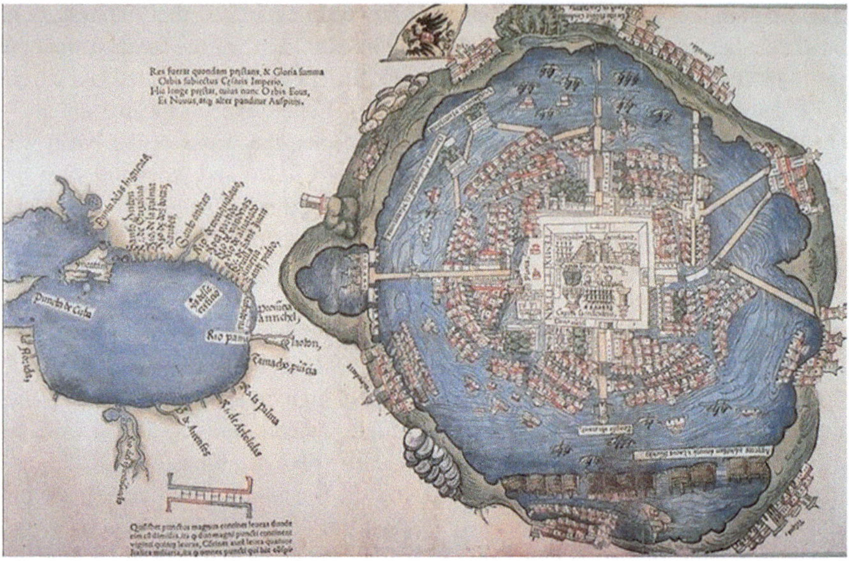

Fig. 1.4 Map of Nuremberg. A representation of Mexico-Tenochtitlan ordered by Cortes. Published in Nuremberg Germany in 1524

Moyotlan, and (4) Teopan. These roads had different functions, such as docks and loading piers for boats that transported goods and wastes of the city, but also to control the fluctuations of water levels. This was described by Cortes's letters where he mentioned: "*At all the entrances to the city, where the canoes are unloaded and where the various products used to feed the inhabitants accumulate, there are huts where the guards stay to collect a contribution for each product. I do not know whether this tax is for the benefit of the emperor or the city, I have not been told, but I believe it is for the benefit of the emperor, because in the markets of other provinces the tax was paid to the lord of the place.*" On the other hand, these causeways had an additional meaning for the Aztecs as they also represented the four cardinal points of the universe according to their mythology (Biar 2020).

By the Spaniards arrival, it was from this city that Moctezuma II, the last Aztec tlatoani (ruler), administered about 200,000 km^2 of territory composed of 400 cities located in 38 different provinces approximately (Biar 2020). According to Hernan Cortes's letters, many people assembled in markets for trading many different goods: "*This city has many public squares, in which are situated the markets and other places for buying and selling. There is one square twice as large as that of the city of Salamanca, surrounded by porticoes, where are daily assembled more than sixty thousand souls, engaged in buying, and selling; and where are found all kinds of merchandise that the world affords, embracing the necessaries of life, as for instance articles of food, as well as jewels of gold and silver, lead, brass, copper, tin, precious stones, bones, shells, snails, and feathers.*" From this and different sources, it is known that Mexico-Tenochtitlan was one of the largest and most important cities in the continent.

1.4 Hydraulic Infrastructure: From Potable Water Supply to Flood Control

One of the most outstanding features of Mexico-Tenochtitlan city was the complex hydraulic infrastructure developed to exploit the complex system of lakes, along with the innovative urban planning of the city to communicate with the surrounding suburban areas. Managing the water resources in the complex hydrological system of the Valley of Mexico was an important task coordinated by Aztec engineers and high-ranked administrators that, as the capital city and suburban areas grew, involved building various complex hydraulic infrastructure including dikes, causeways, bridges, floodgates, and aqueducts (Luna Golya 2014).

1.4.1 Potable Water Distribution

As for any civilization, an essential task for the Aztecs was ensuring a constant potable water supply for the entire population. Mexico-Tenochtitlan had different sources of fresh water including some springs, such as the sacred ceremonial spring of *Tozplatl,* and the surrounding lakes. Initially, the potable water was transported in mud containers called *cantaros* to then storage it in larger reservoirs (Becerril and Jiménez 2007). This was the only way to transport water until a first aqueduct was built in 1416 to more efficiently manage and transport this potable water, the *Chapultepec aqueduct* (Becerril and Jiménez 2007; Sosa-Rodriguez 2010). This aqueduct, also named the *great aqueduct*, was transporting the water from the Chapultepec hill spring to the city (Raynal-Villasenor 2020). As described by Palerm (1973), this aqueduct consisted of two main pipes built with compacted soil and wood. The main pipe was used for daily water distribution and the other served as replacement when the main pipe was being repaired or cleaned. This was also described by Hernan Cortes in 1520: "*Along one of the causeways to this great city run two aqueducts made of mortar. Each one is two paces wide and some six feet deep, and along one of them a stream of very good fresh water, as wide as a man's body, flows into the heart of the city and from this they all drink. The other, which is empty, is used when they wish to clean the first channel. Where the aqueducts cross the bridges, the water passes along some channels which are as wide as an ox; and so they serve the whole city. Canoes paddle through all the streets selling the water; they take it from the aqueduct by placing the canoes beneath the bridges where those channels are, and on top there are men who fill the canoes and are paid for their work.*"

This project, carried out by one of the most celebrated Aztec engineers Nezahualcoyotl, Texcoco's *tlatoani*, was 12 km long and 7 m wide. As described by Hernan Cortes, the water was distributed through sources and pools located around the city, as well as by canoes paddling through the streets (Becerril and Jiménez 2007). Yet, this structure was not perfect, the city's dependence on water sources that were far from the city increased, and the need of constant maintenance to clean and repair the aqueducts often resulted in water shortages (Sosa-Rodriguez 2010). Since its first construction, this aqueduct was rebuilt and improved over time, one important reconstruction was in AD 1465 (Mays and Gorokhovich 2010). A schematic representation of the rebuilt Chapultepec aqueduct is presented in Fig. 1.5.

As observed in the cross section, the reconstructed aqueduct consisted of a trapezoidal structure of about 1.6 m high and 2.5–3 m wide (top and bottom, respectively) built with masonry and mortar made of lime (Mays and Gorokhovich 2010). This structure laid over a foundation of sand, lime and small rocks along with different pilings to stabilize and support the structure (Doolittle 2011). The Chapultepec aqueduct continued to be repaired after different events that damaged its structure, even after the Spaniards arrival, making it one of the Aztecs infrastructures that can still be seen in present-day Mexico City (Raynal-Villasenor 2020).

In addition to its transportations and distribution, the quality of the water was also monitored by the Aztec rulers, who implemented severe punishments to the people that discarded waste into the lakes or channels around the capital (Tortolero 2000). Sanitary measures were of great importance for the Aztec populations. Special canoes around the city were used to dispose wastes including feces that were collected to fertilize and improve the soils of their agricultural fields, and urine was also reserved in containers to be later used as a fabrics dye (Becerril and Jiménez 2007). Personal hygiene was also an important and common practice for the Aztecs. The rulers of

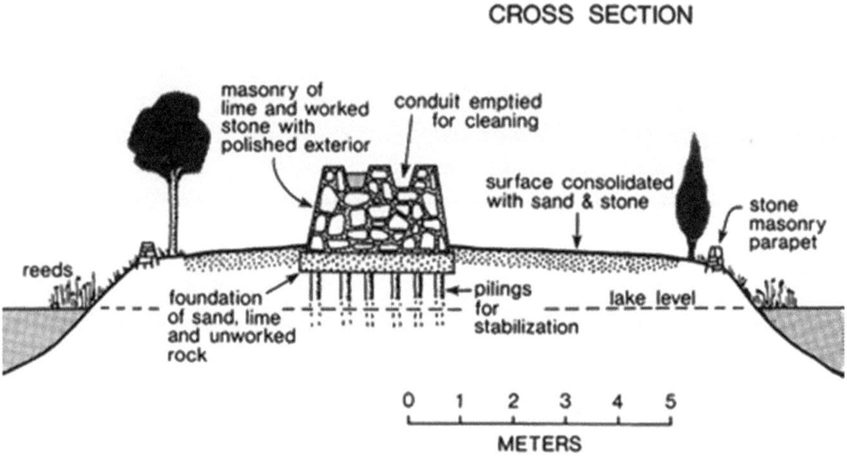

Fig. 1.5 Schematic cross section of the reconstructed Chapultepec aqueduct. Reprinted from Doolittle (2011), after (Bribiesca Castrejón 1958)

the time had their own personal baths, and for the rest of the habitants, public pools were located around the city in addition to steam baths called *temazcales*, a common practice for the Aztec population (Schendel 2021).

1.4.2 Flooding Control

As previously described, the region's lakes were all located at different altitudes and their depth fluctuated with the seasons. Different studies have shown that the depth of Lake Texcoco varied from 0.5 to up to 7 m (Biar 2020; Niederberger Betton 1987; Orozco 1864; Serra Puche 1988). These changes caused great impacts and damages to the city's infrastructure, including the Chapultepec aqueduct, as well as the following reconstructed aqueducts (Ezcurra 1990). Thus, according to Codex Mendoza (1992), using wood, stones, and clay, a first dike was built in the 1400s to protect the city against the water-level changes of Lake Texcoco (Tena 2000). This larger and more complex infrastructure, known as the *Nezahualcoyotl dyke* (see Fig. 1.6), was undertaken to protect the city from floods and it has been recognized as one of the most remarkable and complex Aztec engineering infrastructures. As for the Chapultepec aqueduct, this project was carried out by the well-known Aztec engineer and *tlatoani* Nezahualcoyotl. This great infrastructure divided Lake Texcoco in two parts, creating Lake Mexico around the capital. As described by Torres-Alves and Morales-Nápoles (2020), this dike was about 16 km long, 8 m high and 3.5 m wide. Its structure was composed of different materials, as Lombardo de Ruiz (1973) and Palerm (1973) described: "*Thick wooden piles were stuck to the ground forming a hollow fence, inside the fence, big rocks and sand were deposited between the piles.*" This is clearly displayed in the schematic representation by Torres-Alves and Morales-Nápoles (2020) in Fig. 1.6. According to historical records, this great structure was able to control flooding from its construction until the Spaniards arrival, making a total of about 70 years of successful use (Raynal-Villasenor 2020; Torres-Alves and Morales-Nápoles 2020).

In addition to this notable dike, different roads were built from the capital to the lakeshore. As previously mentioned, these roads allowed communicating more efficiently with the suburban town as canoes were able to circulate through them thanks to different gates located along the roads. Yet, thanks to these gates, these roads also allowed control of the fluctuating water levels (Raynal-Villasenor 1987). Some of these roads can be observed in Fig. 1.4 connecting the city with the mainland. Some of the main roads included the road Tacuba, Tlatelolco-Azcapotzalco, Coyoacan-Xochimilco, Cuitlahuac, and Mexicaltzingo. Additionally, these flow-control features allowed improving fishing activities for the Aztecs as the water flow between lakes was now limited and managed (Raynal-Villasenor 2020).

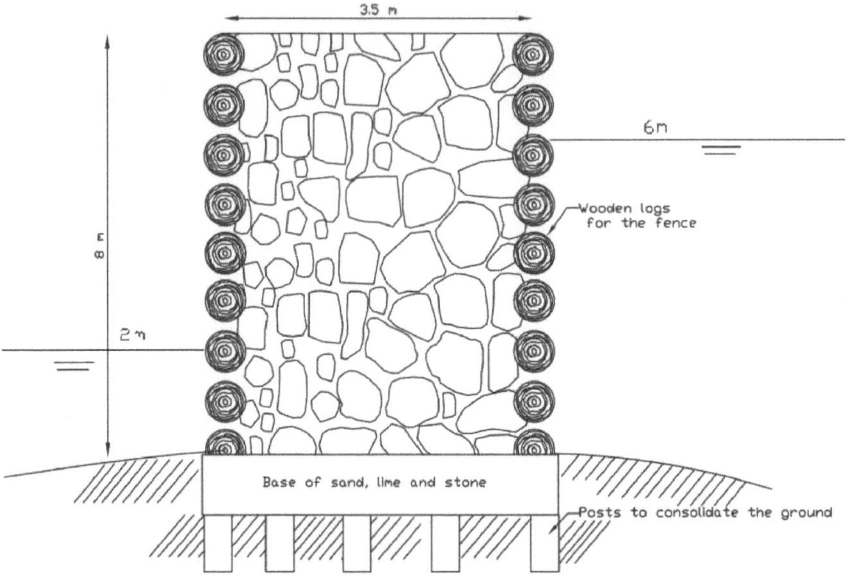

Fig. 1.6 Schematic cross section of the Nezahualcoyotl dike. Reprinted from Torres-Alves and Morales-Nápoles (2020)

1.5 Chinampas: The Innovative Farming Technique

As for all great civilizations, agriculture was one of the backbones of the Aztecs's development. Thanks to its geographic location and innovative engineering, the Aztecs' agriculture was based on *chinampas*, a unique agriculture system often referred to as "*floating gardens*" (Becerril and Jiménez 2007). According to Molina (2010), this term derives from *chinamitl,* a Nahuatl term meaning an area enclosed by canes or a hedge, and has however become to make reference to agricultural region in a wetland with elevated plots surrounded by canals (Morehart 2012). This system is known as one of the most productive and sustainable systems ever developed (Altieri and Koohafkan 2004; Ebel 2020).

Chinampas systems are biodiverse systems with renewable soils that can be used for continuous cultivation and create micro-environments that allow protecting the crops (Morehart and Frederick 2014). Different studies have described chinampas structure and construction as follows (Luna Golya 2014; Robles et al. 2019). First, shallow areas are located by a series of stick soundings. Then, as represented in Fig. 1.7, rectangular "floating" areas are staked out. Vegetation mats, sometimes alternated with lake mud layers, are piled up in the previously delimited rectangular areas. These constructed foundations are capped with thicker layers of mud sometimes extracted from the lake bottom, alluvial sediments or from older chinampas. The chinampas borders are planted with *ahuejotes,* willow trees that grow straight and strengthen the structure from erosion. Finally, a layer of organic muck is added

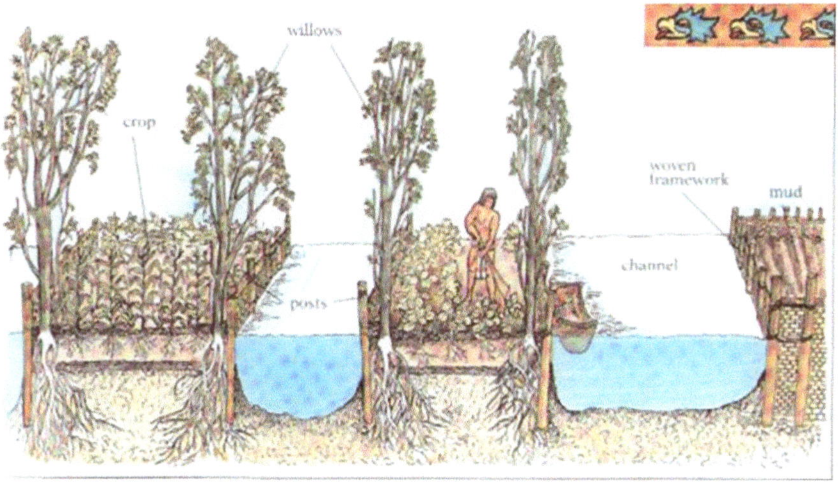

Fig. 1.7 Chinampa structure representation from an unidentified author

before the last process of sowing the selected crops. In average, chinampas consisted of platforms 6–10 m wide and 0.5–0.7 m above the water level.

The availability of water through the surrounding channels provided continuous water to feed the roots of the cultivated plantations. As described in previous sections, the region is known for having irregular precipitation seasons. Thus, this process ensured a continuous irrigation regardless of the season of the year, as the water was constantly absorbed by capillarity when needed.

Similar and comparable systems have been observed in other regions of the American continent. As described by Morehart (2012), raised-field systems have been documented in other Mesoamerican regions such as the Mexican Gulf Coast, northern Belize, Guatemala, (Doolittle 2011; Pohl 2019; Scarborough 2003; Siemens 1983; Whitmore et al. 2001; Wilken 1990) as well as southern American regions including Andean highlands and Amazonian lowlands (Bandy 2005; Denevan 2001; Walker 2011; Zimmerer 1991). However, this innovative agricultural system has been certainly a benchmark of Aztec technology, as it is a system that survived the colonial period and continues to be used nowadays.

As observed in Fig. 1.8, chinampas can still be found in Mexico City. Current chinampas are generally privately owned land with average areas of 1,400 m^2 and altitude of 2240 m in Mexico City's southern Lake Xochimilco (Ebel 2020; Merlín-Uribe et al. 2013). Chinampas keep playing an important role in the economy, environment and culture of the region by reintegrating pre-colonial traditions and technology (Ebel 2020; Losada et al. 1998). Moreover, chinampas are becoming an innovative effort to reduce greenhouse gas emissions in Mexico City, as these systems have shown to storage large quantities of carbon (Renard et al. 2012).

Fig. 1.8 View of chinampas in Mexico City in November 2019. *Photo* Mariana Castaneda-Gonzalez

1.6 Conclusions

Water resources management has been and continues to be one of the essential backbones of progress and stability in past and modern societies. The observed water-oriented culture of the Aztecs highlights how main water resources concepts, such as fresh water availability and flooding protection, have been innovatively treated and solved hundreds of years ago. Through this work, it is possible to analyze how these technologies evolved and sometimes disappeared. Yet, this analysis also allows us to evaluate the possibility of reintegrating them in present-day practices. In other words, this study presented technology from one pre-Hispanic Latin American civilization to highlight past approaches and technologies that not only were successful, but that also created one of the world's greatest civilizations.

Nowadays, supplying water to the world's populations and controlling the impacts of floods and droughts continues to be an issue. This issue has been certainly underlined with the great potential impacts of climate change on water availability in different regions of the world (Konapala et al. 2020). Thus, improved water management strategies that view water as vulnerable and finite resource are certainly of need.

Water management strategies are often defined and identified depending on each particular region and its local needs, so there is no global consensus on which are the best water management strategies. Yet, comparing and assessing the innovative practices observed around the world can shed some light on other future possibilities. Learning from past successful strategies or failed attempts can help create new and improved strategies. The great lacustrine empire of the Aztecs and their technological achievements in agriculture and hydraulic infrastructure are only a few examples of the innovative ancient technologies that can be found in Latin America, and that can be used to reflect and learn to improve water management strategies for the future.

References

Altieri MA, Koohafkan P (2004) Globally important ingenious agricultural heritage systems (GIAHS): extent, significance, and implications for development. In: Proceedings of the second international workshop and steering committee meeting for the globally important agricultural heritage systems (GIAHS) project. FAO, Rome, Italy, pp 7–9

Bandy MS (2005) Energetic efficiency and political expediency in Titicaca Basin raised field agriculture. J Anthropol Archaeol 24:271–296

Becerril JE, Jiménez B (2007) Potable water and sanitation in Tenochtitlan: Aztec culture. Water Sci Technol Water Supply 7:147–154

Biar A (2020) A lacustrine cultural landscape in the prehispanic Basin of Mexico. Int J Naut Archaeol 49:341–356

Bribiesca Castrejón JL (1958) El agua potable en la República Mexicana; los abastecimientos en la época prehispánica

Codex Mendoza (1992) Berdan FF, Anawalt PR (eds) University of California Press

Denevan WM (2001) Cultivated landscapes of native Amazonia and the Andes. Oxford University Press, USA

Doolittle WE (2011) Canal irrigation in prehistoric Mexico: the sequence of technological change. University of Texas Press

Ebel R (2020) Chinampas: an urban farming model of the Aztecs and a potential solution for modern megalopolis. HortTechnology Hortte 30:13–19

Ezcurra E (1990) De las chinampas a la megalópolis: el medio ambiente en la cuenca de México

Konapala G, Mishra AK, Wada Y, Mann ME (2020) Climate change will affect global water availability through compounding changes in seasonal precipitation and evaporation. Nat Commun 11:3044

Legorreta Gutiérrez J (2006) El agua y la Ciudad de México: de Tenochtitlán a la megalópolis del siglo XXI. Universidad Autónoma Metropolitana, Unidad Azcapotzalco, División de

Lombardo de Ruiz S (1973) Desarrollo urbano de México-Tenochtitlan según las fuentes históricas

López Luján L (2002) La cuenca de México durante la época mexica. Atlas histórico de Mesoamérica 148–152

Losada H, Martínez H, Vieyra J, Pealing R, Zavala R, Cortés J (1998) Urban agriculture in the metropolitan zone of Mexico City: changes over time in urban, suburban and peri-urban areas. Environ Urban 10:37–54

Luna Golya GG (2014) Modeling the aztec agricultural waterscape of lake xochimilco: a gis analysis of lakebed chinampas and settlement

Mays LW, Gorokhovich Y (2010) Water technology in the ancient American societies. In: Mays L (ed) Ancient water technologies. Springer, Netherlands, Dordrecht, pp 171–200

Merlín-Uribe Y, González-Esquivel CE, Contreras-Hernández A, Zambrano L, Moreno-Casasola P, Astier M (2013) Environmental and socio-economic sustainability of chinampas (raised beds) in Xochimilco, Mexico City. Int J Agric Sustain 11:216–233

Molina Ad (2010) Vocabulario en lengua castellana y mexicana

Morehart CT (2012) Mapping ancient chinampa landscapes in the Basin of Mexico: a remote sensing and GIS approach. J Archaeol Sci 39:2541–2551

Morehart CT, Frederick C (2014) The chronology and collapse of pre-Aztec raised field (chinampa) agriculture in the northern Basin of Mexico. Antiquity 88:531–548

Mundy BE (2021) The death of Aztec Tenochtitlan, the life of Mexico City. University of Texas Press

Musset A (1989) L'eau dans la vallée de Mexico: enjeux techniques et culturels (XVIe-XIXe siècle). EHESS, Paris
Niederberger Betton C (1987) Paleopaysages et archeologie pre-urbaine du Bassin de Mexico (Mexique)
Orozco M (1864) Memoria para la carta hidrográfica del Valle de México. A. Boix
Palerm Á (1973) Obras hidráulicas prehispánicas en el sistema lacustre del Valle de México
Pohl MD (2019) Ancient Maya wetland agriculture: excavations on Albion Island. Routledge, Northern Belize
Raynal-Villasenor JA (1987) The remarkable hydrological works of the Aztec civilization. Water for the future: hydrology in perspective. Int Assoc Hydrol Sci 164:3–9
Raynal-Villasenor JA (2020) Hydrologic and hydraulic works hydraulic works of the Aztec civilization Aztec civilization. In: Raynal-Villasenor JA (ed) Water resources of Mexico. Springer International Publishing, Cham, pp 237–254
Renard D, Iriarte J, Birk J, Rostain S, Glaser B, McKey D (2012) Ecological engineers ahead of their time: the functioning of pre-Columbian raised-field agriculture and its potential contributions to sustainability today. Ecol Eng 45:30–44
Revueltas Valle E (2017) El paisaje cultural de Tlatelolco, en el códice Ordenanza del Señor Cuauhtémoc, 1521–1527, Universidad Autónoma Metropolitana (México). Unidad Azcapotzalco
Robles B, Flores J, Martínez JL, Herrera P (2019) The Chinampa: an ancient Mexican sub-irrigation system. Irrig Drain 68:115–122
Scarborough VL (2003) The flow of power: ancient water systems and landscapes. School for Advanced Research on the
Schendel G (2021) 3. Aztec Public Health and Welfare. Medicine in Mexico. University of Texas Press, pp 32–44
Serra Puche MC (1988) Los recursos lacustres de la cuenca de México durante el formativo. Unam
Siemens AH (1983) Wetland agriculture in pre-Hispanic Mesoamerica. Geog Rev 166–181
Sosa-Rodriguez FS (2010) Impacts of water-management decisions on the survival of a city: from ancient Tenochtitlan to modern Mexico City. Int J Water Resour Dev 26:675–687
Tena R (2000) Ordenanza del señor Cuauhtémoc. MAG Ediciones
Torres-Alves GA, Morales-Nápoles O (2020) Reliability analysis of flood defenses: the case of the Nezahualcoyotl dike in the Aztec city of Tenochtitlan. Reliab Eng Syst Saf 203:107057
Tortolero A (2000) El agua y su historia: México y sus desafíos hacia el siglo XXI. Siglo XXI
Walker JH (2011) Social implications from agricultural taskscapes in the southwestern Amazon. Lat Am Antiq 22:275–295
Whitmore TM, Whitmore TM, Turner BL (2001) Cultivated landscapes of Middle America on the eve of conquest. Courier Corporation
Wilken GC (1990) Good farmers: traditional agricultural resource management in Mexico and Central America. University of California Press
Zimmerer KS (1991) Wetland production and smallholder persistence: agricultural change in a highland Peruvian region. Ann Assoc Am Geogr 81:443–463

Chapter 2
Water in Inca Cosmology and Statecraft

Jeanette E. Sherbondy

Abstract This chapter discusses the central importance of water in Inca cosmology. Their awareness that water is life frames their hydrological cosmology as a circulation system that involves the interior of the earth, its surface, and the spaces above it with a cosmic ocean underlying everything. This cosmology evolved as a branch of the cultural development of the peoples who inhabited the Lake Titicaca basin for millennia reaching its apex in the Tiahuanaco (Tiwanaku) state (300 AD-1100 AD), the earliest and most powerful polity on the Altiplano (Berenguer in Tiwanaku: Lords of the Sacred Lake. Museo Chileno de Arte Precolombino, 2000; Janusek in J Archaeol Res 12, 2004; Plourde and Stanish in The Emergence of Complex Society in the Titicaca Basin: The View from the North, 2005; Stanish 2013). Lake Titicaca was revered as the place of origin of the ancestors of the Incas and they saw their own importance as a product of their origins in the lake. Water concerns were central to Inca history and policy as their initial decision to settle in Cuzco was based on the quality of the water and soils there. Subsequently, they organized the landscape and the population for the best management of waters in the Huatanay river drainage basin, where Cuzco is located. Water concerns regulated their ritual and agricultural calendar as it was their mode for maintaining the equilibrium of the universe.

Keywords Inca · Cuzco · Lake Titicaca · Tiahuanaco · Wiracocha · Water · Cosmology · Religion · Hydrology · Puquina

2.1 The Incas

The Inca empire was the largest pre-Hispanic empire, founded at Cuzco, its capital in the upper Huatanay River valley. This valley and the nearby Urubamba offered ideal environments for cultivating maize, complemented with higher lands used for growing potatoes and other tubers and raising llamas and alpacas. The Huatanay valley was already populated by other small polities in the 14th and early fifteenth

J. E. Sherbondy (✉)
Washington College, Chestertown, MD, USA
e-mail: jsherbondy2@washcoll.edu

G. A. Corzo Perez et al. (eds.), *Aqua Vitae*, The Latin American Studies Book Series,
https://doi.org/10.1007/978-3-031-57731-4_2

centuries, who spoke Puquina and Aymara languages. The Incas were not the original people but immigrants who conquered the other polities and extended their control to the Urubamba valley to the northeast, proceeded to conquer the rest of the Huatanay valley to the southeast, incorporating the earlier Wari city known as Pikuillacta (Glowacki 2012; McEwan 2005).

The Incas professed their own origins were in primordial times emerging from the waters of Lake Titicaca (Molina 1943[1583]). They put themselves in the context of the myth of Wiracocha, a deity who emerged at a time of darkness out of the lake. He brought the sun out first of all and placed it where it belonged, in the sky. Then he placed the Moon and stars. The Incas claimed they had emerged right after the sun, making them children of the sun. Wiracocha commanded them to resubmerge themselves to follow underground channels until they reached the Cuzco area, where they appeared on the earth's surface at Pacaritambo, a few kilometers south of Cuzco. In their myths, they came out with all their royal trappings, thus defining themselves as superior to all others, by linking themselves to the lake as a privileged family from the original times of the present earth. They were the Ayar family, who were speakers of Puquina, the language of Tihuanaco empire, and Aymara, the language that succeeded Puquina in the Titicaca basin by AD 1100 (Cerrón-Palomino 2013, 2016). Their origins in Lake Titicaca sanctioned by Wiracocha were essential to their claim of superiority.

2.2 Inca Religion

The worship of Wiracocha and the sun at Tiahuanaco became the basis for their own Inca religion. By the time the Spanish arrived, the meaning of "Wiracocha" had evolved to mean, for some of the Quechua speakers, the "foam of the sea" but its linguistic archeology reveals an Aymara origin (Cerrón-Palomino 2013). The full name of the deity as conserved by Inca royalty (Betanzos 1996[1551]) was Contiti Wiracocha. Contiti was the earlier name in Puquina dating to the Tiahuanaco state, "titi" meaning sun and "con" meaning burning hot.

The invasion of the Titicaca basin by Aymara-speaking chiefdoms with the dissolution of the Tiahuanaco empire involved replacing Puquina with Aymara. Subsequently, they forgot the meaning of Contiti, the burning hot sun god and so they referred to him as "Wiracocha" meaning "he who has the bloody lake" (Cerrón-Palomino 2013). The name Contiti was preserved at least among the priestly elite, along with the association with Lake Titicaca, where blood sacrifices (mostly of llamas) were made to the water. The later Quechua speakers who did not know Aymara were left to make their own guesses about the meaning of "Wiracocha" to come up with "foam of the sea," which is not grammatical Quechua (Cieza 1959[1553]). The Quechua term Pachayachachi was added to the other two names when that language became the official one only a few generations before the conquest. It means "he who put the world in order" or "hacedor." That change, in turn, led the Spanish clergy to reinvent him as the Christian creator deity (Betanzos 1996[1551]; Cerrón-Palomino; Dumezil).

The Incas repeatedly attacked the Collao kingdom on the northern shore of the lake that still spoke Puquina and eventually conquered it—a sort of reconquest. They ritually celebrated their lake origins in their story of king Inca Roca who was anointed with water carried all the way from the lake to Cuzco in a precious vial. Inca kings continued to travel to the lake to make sacrifices in times of dire need (Cobo 1990[1653], Sarmiento).

Among the several versions of the Incas' initial arrival at Cuzco, is a narrative of their migration to Cuzco from the Collao written by Joan de Santa Cruz Pachacuti Yamqui Salcamaygua (PYS) who composed his "Relación de antigüedades deste reyno del Piru" in 1613, 80 years after the invasion of the Inca empire. He presents a unique viewpoint from the cultural region of his birth located about halfway between Cuzco and Lake Titicaca, a region that had been Puquina-speaking, then Aymara-speaking and was on the main road that linked Tiahuanaco to the Huatanay valley. Other versions of the Incas' trek to Cuzco start at Paccaritambo, a site only a few kilometers south of Cuzco, but PSY incorporates this segment into a larger narrative linking the Incas to the Collao and to three kin groups: their father's, their father's brothers', and their mother's brothers'. This detail makes their history more realistic, whether it was true or not.

PYS attributes the Inca diaspora to a drought in the Collao that led to a struggle with other chiefdoms for lands with reliable sources of water. Tunupa, the lightning bolt deity of the Lake Titicaca basin honored Apu Tambo, father of the Ayar siblings who were the first Incas, by giving Apu Tambo his staff of authority with his blessing. This staff turned to gold when Manco Capac, the eldest of the four sons was born. After his father and mother died Manco Capac, who became chief, reviewed the remaining subjects of his father to see if they were up for a new conquest (PYS). He and his seven siblings led those who were willing, to Pacaritambo, where they emerged from under the earth.

From there, they proceeded toward Cuzco to the north carrying Tunupa's golden staff to the Huanacauri, a very large mountain that dominates the Cuzco valley in the south. "Huanacauri" means "rainbow" in Puquina. The rainbow is intimately associated with the Amaru serpent, which represents it in the form of a double-headed arc (Urton 1981a, b). Huanacauri, the mountain, was the source of many irrigation canals during the Inca period. Its subterranean capacity for storing water is enough to provide for 88 years of drought (p.c. by the on-site archeologist Milthon Maque Ramos (2021).

After the Incas conquered the wak'a,[1] Huanacauri, Manco Capac, and his sister Mama Huaco entered the Huatanay valley and tested the soils in various spots by inserting Tunupa's staff into the ground (PYS, Sherbondy 1982a). The place they chose was where they found soils, that were neither loose gravel nor compact clay, but probably swampy soil that is rich in nitrates and abundant water (Sherbondy 1979a,

[1] 1 "Wak'a" is another word inherited from the Puquina language. It means something sacred, frequently a topographic feature, anthropomorphized, and represented by a stone (CP, p.c. 2022).

1982a). The Incas located their Cuzco settlement between the two major streams that form the river. Under the auspices of Tunupa, the Incas had found the "good water" for which they were searching.

Where the two rivers joined, there was a field named Sausero that was incorporated into Inca ritual as the place where the Inca plowed up the soil for the first time each spring, marking the time to begin to sow maize. Near it the Incas built the temple dedicated to Tunupa, called Kuri-n(a) Kachi, the original Puquina for that later evolved into Coricancha (Bouysse-Cassagne 1993; CP 2016). Much later the name was remodeled in Quechua either "deliberately or spontaneously" as < Cori cancha > and it took on the meaning of temple of the sun (CP 2016). This linguistic development underlines the original emphasis on the source of water as rain in the form of the god of lightning and thunder rather than the sun. Even though the Inca myths collected in the sixteenth century emphasized that it was first and foremost the temple of the sun and of the ancestors, where their mummy bundles were installed, this was not the original plan.[2]

Cuzco Valley showing the temples and plazas that were the nucleus of Inca Cuzco (upper left). The Huatanay River valley at its source on the left has a gap where it flows to the south. Huanacauri mountain on the lower right.

Manco Capac's son and successor as Inca Lord made a political marriage to the daughter of the chief who controlled the lower part of the Huatanay valley. This part of the valley was originally marshy land and Sinchi Roca was credited with building a drainage system of the meandering river to create more cropland. This brief outline demonstrates how the Inca rulers acquired strategic control over all the water sources and the whole river bottom that form the Huatanay River (Sherbondy 1982a).

The Temples in Cuzco of the Major Inca Deities

In myth, religion, and policy, the early Incas were dedicated to control over water. They also identified themselves initially with the ancient Tiahuanaco deity of lightning who causes rain, Tunupa (Puquina name). Tunupa became Illapa under the Incas. Illapa was imagined as a man who lives in the sky with the stars. He carries a war club in his left hand and a sling in his right hand. He wears shining garments that give off flashes of lightning when he whistles his sling. The crack of the sling is thunder when he wants it to rain. He crosses the large river in the sky (the Milky Way) and pulls water from it to let it fall to earth in many forms, as rain, clouds, and hail. He is the embodiment of thunderbolts, lightning, rain, hail, and even storms and whirlwinds (Cobo 1990[1653]). His Inca name, Chuqui Illa, refers to the light he radiates as Thunder, the "bright glow of gold."[3] The essence of power of the two

[2] Cerrón-Palomino (2016) introduced the term "linguistic archeology" for his research on Andean languages. His research on the original locations and times of the origins of Puquina, Aymara, and Quechua, has contributed to important changes for Andean chronology. Furthermore, his research into the hermeneutics of place names, proper names, topographical names, names of institutions and concepts is clearing up some confusions in the scholarly literature about Inca history and culture as well as introducing new and challenging perspectives.

[3] Cobo described systematically the groups of major forces, i.e., "deities" in pp. 22–34.

deities, Illa and Inti (lightning and sun), was in the light that both gave off, their radiance, their glow, their shining (Cobo 1990[1653]).

The Incas also emphasized their superior rank among human beings as the children of the sun, Titi, whom they renamed Inti, as a way to link the sun to themselves as coming from the Anti, the southeastern direction (Cerrón-Palomino). Wiracocha had become the Lord of the Bloody Lake before the Incas arrived in Cuzco losing the earlier Puquina association with burning hot sun, Contiti, even though the memory of Wiracocha as Contiti persisted among the priests (Betanzos 1996[1551]; Cerrón-Palomino). In Cuzco, his temple was constructed on the current site of the cathedral where water from springs on the mountain slope pooled. This is an area of many springs and streams that flow toward Cuzco (Sherbondy 1982a). At the base of the slope in front of Wiracocha's temple, there was a large sunken plaza where water filtrating from the slope collected. This plaza, now the Plaza de Armas, was known as the Aucay Pata, meaning in Puquina the "Big Plaza" (CP??). On the west side, the Saphi River flowed separating the Big Plaza from another smaller one known as Cusipata. Together these plazas formed a large open space at the center of the city with the Saphi River flowing between the two plazas.

The Incas filled Aucay Pata with sand from the Pacific beaches to a depth of two and a half hands and more in some places according to the Spanish governor, Polo de Ondegardo. His Inca informants declared they brought the sand from the sea to give reverence to "Ticsi Viracocha" as the deity of the sea, confusing < Ticsi > for the earlier < Titi >." They buried in the sand vessels of gold and silver as well as figures of llamas, alpacas, and human beings.[4] Polo immediately ordered that the sand be removed. He had it used in the construction of the cathedral to subordinate Viracocha to the Christian god (Sherbondy 1992). They also built their sacred Sunturwasi and Ushnu, a sort of pyramid with a hole that penetrated the earth and where the Inca king made sacrificial offerings to the watery interior.

The idea that Cuzco was built on a lake has persisted until the present. When I was in Cuzco in the mid-1970s, people firmly believed that golden ducks swam on this subterranean lake and pointed to the Unu Punku chapel in the cathedral as the door to the subterranean lake (1982a; Dominguez Faura 2017).

The temple to the lightning, Illapa [Tunupa] was built behind it. Sacsahuaman, considered to be a fortress because of its zig zag walls that reminded the Spaniards of their fortifications, was originally a great temple to lightning and thunder (Barreda Murillo 1973).

The Incas developed their sun deity Inti into three forms: one for the sun at dawn, one for noonday, and one for the late afternoon setting sun, Inti (CP 2016). This Inti was the mature Lord Sun who was married to the Moon his wife and who had the stars as his children. Thunder also had three forms and names. They were identified by kin relationship, he himself, his brother, and his son. Each of the three suns and three thunders had a textile bundle which flanked the Wiracocha bundle on either side according to the report in 1653 by Bernabé Cobo (1990).

---

[4] Polo de Ondegardo (1917 [1571]) cited by Sherbondy (1982a, 1996).

The relationship of thunder to sun is oppositional as well as complementary. If the sun shone all the time, there would be no rain, and drought would endanger all things. Thunder and Lightning need to bring rain, but only for a time, to prevent crops from rotting. This dynamic is a source of existential tension in the highlands. The myth of Wiracocha seems to reconcile these two powerful deities in the form of alternating seasons.

Several places were made wa'as to Mama Cocha inside of Cuzco (Cobo 1990[1653], and the lake Coricocha, on top of the mountain NNE of Cuzco, that is still considered the "Mama Cocha" (Sherbondy 1982a, b), so even though the Spaniards did not report any other major special temple to Mama Cocha, she was one of the major deities. Female deities and religious rituals were generally not communicated to the male Spanish clergy (Silverblatt 1987).

2.3 The Spatial Organization of Inca Cuzco

The Inca used several methods for indicating locations on the topography of Cuzco that are usually seen in small villages but were developed into a complex map by the Incas. The base line literally was the NW–SE axis that follows the natural fault lines of the river in the Cuzco valley. It is also found in the line that divided the Lake Titicaca basin; a line referred to as the aquatic axis (Wachtel 1990; Bouysse-Cassagne 1987). Urton (2021) argues that this is a mirror reflection of the celestial axis of the river (Milky Way) that is foremost in their astronomy. Even the smallest villages were familiar with the motions of the Mayu in the night skies. The fault lines that run from the Titicaca basin to the Cuzco valley were formed by the uplift of the Nazca plate under the South American continent.

In Lake Titicaca, the two halves were Urco Suyu, the Mountain side on the west, and Oma Suyu, the Water side on the east (Bouysse-Cassagne 1987). In Cuzco, the Incas called the northern side Hanan or the Upper side while the Lurin (Hurin) designated the Lower.[5] These dual divisions followed geographical features in both cases. The Incas proceeded to divide each one in two again, resulting in a four-part division of suyus, that became the designation of their empire as the Tahuatinsuyu, meaning the four suyus as a whole.

One of the functions of these divisions was the organization of the lands and waters that the royal Inca descent groups ("panaca" in Puquina) as well as some privileged non-royal kin groups (ayllus in quechua). There were ten official panacas headed by each Inca king, five in Hanan Cuzco and five in Lurin Cuzco that were ranked in order of the chronology of the reigns of the kings. There were also ten ayllus. They were divided among the four suyus that had been further divided into 41 sectors by lines called ceques that radiated from the central point. Most of the ceques ended at the point where the drainage basin ends, although there were exceptions in the case of irrigation canals that brought water to Cuzco from another drainage basin.

[5] CP argues that "Lurin" or "Urin" are preferable spellings, not "Hurin."

The line was extended to include the sources of the canals so that the effective limits marked the de facto drainage basin, whether by natural streams or canals (1982). There is an indication that the Incas were dividing Contisuyu into two, anticipating a five-part division when the later Incas introduced a decimal mathematical system (Urton).

This radial map of Cuzco was based on the center, the Coricocha temple. Villages with a simplified radial internal division are still common in the area south of Cuzco, incorporating the ancient territory of the Tiahuanaco state, such as San Andrés de Machaca, near the ruins of Tiahuanaco (Albó 1972; Sherbondy 1982a).

In general, the suyus were subdivided into groups of three ceques to form about three sectors that were ranked hierarchically as collana (upper), payan (middle), and cayao (lower), that is often used to organize work groups. Each line was assigned to a panaca or ayllu, some to two ayllus. Panacas generally had two or three sectors.

A ceque line assigned to one panaca had anywhere from 4 to 16 wak'as or sacred places on it. Imagine it as knots on one string of a khipu, knotted string apparatus that the Incas used for keeping accounts (Urton). This is the "to-do" list for each descent group that was granted a place in the valley. It gave them a sociopolitical rank among the twenty-some panacas and ayllus and along with it, concrete obligations to the Inca state for providing the requisite sacrifices and offerings. Of the 41 ceques and 328 wak'as at least a third are either sources of water or are significantly associated with water (Sherbondy 1982a).

The most basic principles about ownership of farmland and rights to water underlie this organization of the land. As in the myth of Wiracocha's ordering of the universe, the lands and waters belonged to the ancestors, sacralizing ownership of water sources and the lands dependent on them. The Inca state placed the panacas and ayllus where they were expected to take on the obligations of making the offerings to their assigned wak'as and of maintaining the infrastructure, and thus granting them the right to work the lands and irrigate with the waters (Sherbondy 1982a).

Inca religion was deeply political. The ceque assignments to descent groups established and made visible their position in the social hierarchy, and their obligations needed to be fulfilled in order to maintain their position. The divisions of lands and waters and ritual duties were the spatial aspect. Temporal control was accomplished by means of a state calendar of festivals and sacrifices.

2.4 The Ritual Calendar: Maize, the State, and the Cosmos

Maize was by far the most important crop for the Incas (Murra 1980). It was the heart of the economic system and facilitated the festivals that were essential for maintaining political power. Maize was the most desired over other foods. Dried maize was used to feed the armies as well as all the retainers of the Incas. It was also dried, ground up, boiled, strained, and allowed to ferment as the everyday and most prized ritual drink.

Irrigation is necessary to grow maize in Cuzco because the natural rainy season is not long enough to support the maturation of kernels before the killing frosts of the dry season (Sherbondy 1982a). The water requirements for maize depend on irrigation for the sowing and early sprouting during the dry season just before the rainy season begins (Sherbondy 1982a).

The annual cycle of festivals and ceremonies also functioned to plead for rain at critical times during the maturation of the maize plants. Fears of droughts and floods were dramatically expressed in their prayers and offerings from August until December. After that, they ritually celebrated the success of the harvested crop until June. The harvest and end of the agricultural year was marked by the festival of Inti Raymi at the June solstice, literally the Sun Festival in Puquina (CP). The following description of the annual agricultural and ritual year is based on Cobo 1990[1653], Guaman Poma de Ayala 1980[1615]) and my research (Sherbondy 1982a).

The next six months are concerned with preparing the land for the new maize crop by irrigating, plowing, planting, hoeing, and praying for rain. Lands were inspected and redistributed by state officials. Maintenance of the irrigation systems and of the terraces was crucial because the rains, the trampling of canals by animals, and land slips, all eroded the canal walls so that every year they needed to be repaired by the users of that canal. Then the first round of irrigation could begin in order to moisten the soil so that a couple of days later the seed corn could be planted (Sherbondy 1982a).

At this time, Tocori puquio was worshiped as the spring that was the principal source of water for Cuzco. Two brown llamas were sacrificed, one at the upper end where the source was and the other one was sacrificed at the last point on the Huatanay River where it flows into the Vilcanota River. This ritual was to preserve the water (Cobo (1990[1653] Chap. 28).

Tocoripuquio (Chacan) is where Inca Roca created a more voluminous water supply for Cuzco according to myth.

> This sacrifice was established by Inca Roca [founder of Vicaquirao panaca], and the Indians tell this fable about him. Very little water used to come out of that spring, but after making certain sacrifices, Inca Roca thrust his arm into the spring, and this made the water gush out as abundantly as it does now. This they considered to be such an established fact that the members of his family unit and lineage used it as justification for laying claim to that water for their own use exclusively, and during the time of the Incas, they succeeded in getting it. (Cobo 1990[1653] Chap. 29).

Inca Roca's panaca was in charge of the ceques that included the wak'as of this water source.

People reserve those land and water rights by their labor. They must cultivate and care for the fields and the same principle applies to water rights. Maintaining the sources and the canals by repairing them every year or whenever necessary confirms their rights to their share of irrigation water. The offerings to the sources of water are also part of their obligations (Sherbondy 1982a). These assumptions continue to the present day, even though the Spaniards introduced the purchase of water rights and land rights (Sherbondy 1979a).

In August after the first round of irrigation, the planting began. The Inca used a foot plow, and his Coya (Queen) sowed the first kernels of maize in the field of Sausero. Dances and songs were their prayers for a good year. Fifteen llamas were distributed for sacrifices at each of the 328 wak'as in the area surrounding Cuzco, that is, all the wak'as named on the ceque system.

The ritual included a white llama, Napa, with its ears of gold, tethered in the middle of the field of Sausero. People fed the llama corn beer and kept on feeding it as long as the sowing lasted. At the end, a thousand guinea pigs from all the provinces were sacrificed to frost, wind, and the sun and to forces that make their crops grow or could harm them. The designated priest for the planting fasted until the maize sprouted and grew one finger above ground. This was the sign that the seed was viable, a visible proof of success.

The Situa ceremony, related to the September equinox, was a massive cleansing ceremony for the Inca to rid the land of harm and illness. After prying for success and abundance during planting, this ceremony concentrated on pleading to Wiracocha not to let illnesses and disasters destroy the crops. The Inca and his lords initiated the Situa at Coricancha temple with a cleansing by fire. As soon as the new moon appeared, they started shouting while carrying torches in their hands, "Sickness, disasters, and misfortune, get out of this land." Everyone came out with torches and did the same (Cobo 1990[1653]).

Four groups of 100 warriors each were organized to chase the bad things out of the city with their spears along the four roads. They gathered in front of Coricancha and from there ran on the four royal roads out to the junctures of those roads with the largest four rivers in each direction. The warriors bathed and washed their clothing and their weapons to rid themselves of all misfortune. Everyone else in Cuzco ran to the ends of their ceques. All four rivers carried the sicknesses out to the Apurimac tributary of the Amazon River that then flowed into the Cocha Mama.

When they returned, they ate a special corn or quinoa pudding mixed with blood and smeared some of it on their faces, doors, and other places where they kept their food and clothing. They also took some to the springs and water sources and fed it to them. It was a period of rejoicing, of no disputes, of eating and drinking the best available fruits of the harvest so that the people would not be plagued with illnesses or bad fortune. After this, everyone from the Inca and his lords to the poorest commoners gathered in the Aucay Pata in front of Wiracocha's temple, with the mummies of their ancestors, to feast with them in hopes for a good year.

October and November are the hot dry months. Light rains usually begin sometime in November. Irrigation cycles may be started again if the rains are late or insufficient. They needed the Uma Raymi (Rain Festival) to bring the needed rain. There was much anxiety about a possible drought. Their worries intensified to the point of hysteria. The ritual consisted of crying and screaming for rain. Men and women cry, children, sick people, blind, crippled, old men and women, everyone who can cries. A black llama is tied up and not given anything to eat or drink and so that it cries. They pray to Wiracocha to send rain. The black llama invokes the black constellation of the Llama in the Mayu that drinks up water from the earth. The purpose of this ritual was to plead the celestial Llama to cry and send rain for their fields.

The Itu ritual was a special one reserved for the worst years when the rains failed entirely. It was an exclusively male ceremony. The men made sacrifices of animals, but they could also include children. It was a silent ritual. A similar ritual offering occurred in 1986 when I was in the field, the incident that Therese Bouysse-Cassagne (1987) refers to, in which one man was sacrificed by his community in a desperate plea for the rains to stop flooding the fields and destroying the crops. There was yearly tension about whether it would rain and it increased until the rains started. Then when they felt relief and confidence, they celebrated with their ancestors in the Aya Marca ritual. The people carried the mummies of their ancestors from house to house on litters and feasted with them.

Capac Raymi, the Great Raymi of the year, was celebrated on the summer solstice in December. This was the time of the year when the rains were most surely to be falling. After a month of celebrating the summer rains, they enjoyed a month of rest during Camay Quilla, in January, the month of the heaviest rainfall. These rituals involved much drinking, a symbolic representation of rain.

In February, they made a series of dams and then let the first one burst open so that water gushed through all of them passing through the streets and out of Cuzco, cleaning it of debris. This was Mayucati. They threw into the water the ashes of the old sacrifices to send them out to the Mama Cocha.

In March, the crops began to mature. In April, they celebrated that there was no more need for rain. They gave the white Napa llama a bowl of beer that it kicked over and spilled, a sign of no more need of rain. Beer could be rain in rituals. Harvesting took place in May and June with much singing and drinking of purple corn beer, purple giving it the color of blood. The Inca officials carried out a food census. The Inti Raymi (Sun Festival) at the June (winter solstice) signaled the end of the maize harvest. It was a time for resting from the harvest work and giving thanksgiving sacrifices to the sun.

The greatest concern of the Incas was the maize crop, and it relied on irrigation at the time of sowing until the rains began. The greatest anxiety was a failed crop due either to drought—the most likely reason—or to floods, hail, or diseases. Their major state rituals were designed to protect the crop at each season. Embedded in the religious calendar were the agricultural tasks as well as rituals with sacrifices to the cosmic forces to bring rain and stop rain. The rituals could be considered a hydraulic technology, vaguely similar to our cloud seeding for rain. In this case instead of changing the chemistry of the clouds, the Incas appealed to the good will of their deities (Sherbondy 1982a). The state controlled the timing of irrigation and sowing as well as the distribution of water and land rights with their wak'as that the designated panacas and ayllus were obligated to make sacrifices to at certain times (Sherbondy 1982a). Zuidema (2010) presented his hypothesis that wak'a had to be recognized each day of the year, so that the ceque system also functioned as a calendar.

This leads to our final consideration of reconstructing the Incas' overall vision of water movements within their universe, their hydrological cosmology.

2.5 The Inca Cosmos

Here, I wish to interpret Inca cosmology as regards their theories of hydraulics. It means explaining the circulation of waters in their universe from their viewpoint.

Joan de Santa Cruz Pachacuti Yamqui Salcamaygua (PYS 1613) made an early graphic attempt to portray the Incas' most sacred forces of the universe as they affected human life. The Catholic Church in 1584 had required that the earlier Inca deities be included in a diagram to show that they were not deities but creations of the Creator. The diagram should show that precisely these things are no longer gods and therefore should not be worshiped any more (Duviols 2008). Therefore, we can assume that the natural components in the diagram were indeed Inca deities. PYS claimed that this was a replica of a wall in the Coricancha temple, but he was born after the conquest and never saw the temple in Inca times.

It shows the main features that pertain to water.

At the top PYS places constellations, individual stars, the sun, and the moon surrounding an oval to represent Wiracocha. Though he does not show the Mayu (Milky Way), he indicates significant features of the sky that are of great astronomical importance for tracking time (the seasons of the year) and space, as it mirrors the aquatic axis of the earth (Urton 1981a, b, 2021, Wachtel 1990). The large oval of Wiracocha is identified by PYS as a large metal late. I would like to suggest that PYS may have had in mind the "bloody lake" that was embedded in the name of Wiracocha and that the ink blot in the center of it may represent Titicaca, the sacred island of the sun rock. Perhaps it is a map of the two most sacred places for the universe: the lake as the place of origins of life and the rock of the sun originally rose to the sky upon Wiracocha's request (Molina 1943[1573]).

On either side of Wiracocha are the sun and the moon as well as other constellations. Below these are placed features of the earth that rest on a base, a grid identified as the storehouses of the wealth of the Incas, Colca pata. All the figures are enclosed in a "temple."

It is not a diagram of the dynamic relationships of all the parts to a whole, except for two clusters near the bottom, one on the left and the other on the right side. On the left, there is the earth with three mountains and the Pilcomayu river flowing away from it. Above the earth and connected to it is a rainbow and along the left side is a lightning bolt labeled Illapa. This cluster shows water in two forms: a rainbow and a river accompanied by lightning which is revered for breaking open the water stored in the sky so that it can fall to the earth. Yet it does not indicate the Milky Way that is believed to be a River of water by Quechua speakers and a Path of lights by Aymara speakers (Arnold). We do not know what the ancient Puquina speakers called it. Furthermore, we do not really know how the people of pre-Hispanic times thought about what the make-up of "stars" was: Sparkling water? Lights similar to lightning?

The overall impression of this part of PYS's cosmogram deals with the sky's sources or indicators of water as rain, though rain is never depicted. Without intending to, PYS shows us the formerly deified natural features that were worshiped in Inca times, which reveal a strong their focus on the sources of water, some of its forms, and its purpose to promote life and abundance.

The cluster of features on the right side is focused on Mama Cocha, meaning the "big lake" in Quechua, Mama Cota in Aymara, but also the "mother lake" which has a stream that runs from the border of the lake to a spring "pucyo."[6] This is consistent with the belief that springs are fed by underground streams from large lakes. Just below the pucyo is a tree labeled "mallqui" which is also considered to arise from the watery underground and was revered as an ancestor. It refers specifically to a cultivated, not wild, tree (Sherbondy 1982b, 1988). PYS places above this cluster a feline, the Andean wildcat that represents the bringer of hail, another form of water; however, it is dreaded for its ability to destroy a crop and kill a baby llama (Urton 1981b).

Above this cluster is a picture of clouds and the label of winter, referring to the annual rainy season between December and March. On the left PYS drew a number of stars and labeled them summer, the dry season when skies are clear and when people hope for rain to plant their crops. Underneath the earth cluster on the left are seven stars, the Pleiades, that signal to farmers when they should plant (Urton 1981a). There is a contrast then of complementary forces for life with sky deities for coming rains in the summer and on the left are the surface deities of the lake, the spring, and the tree that grow from subterranean moisture.

We can distinguish various patterns of thought about the natural world that were revered in Inca religion. The drawing as a whole identifies the sources of Inca wealth in the universe that guided their lives. The astronomical features that produced day and night and the seasons of the year promoted human life as farmers and pastoralists who were dependent on sources of water. There are the indications in the sky of the dry and rainy seasons and below these are two clusters of drawings that illustrate the sacred sources of water from the sky on the left and earthly sources of water that emerge from the underground on the right. The bottom are the storehouses, Colca Pata, where they kept their accumulated wealth. We know from Murra (1980) that these were products of agriculture and herding, such as fine textiles and woven bags of maize that the Incas gave as gifts to ensure, in reciprocity, the loyalty of their subjects (Murra 1980).

Below is my hydrological cosmogram, a composite of features Urton included in his drawing and emphasis on hydraulic features that need to be included for an understanding of the hydraulic cosmology of the Incas.

This sketch of water circulation includes water from above, on the surface, inside, and underneath the earth.

In this cosmogram, water was the base for the earth and is pictured in the cosmograms of Earls and Silverblatt (1976), Urton(1981a, b, 2021), and in my research (Sherbondy 1982a, 1982b). Urton dramatically portrays the central importance of the celestial Mayu, the major circulating source of water that encompasses the earth and replenishes its waters from the cosmic ocean by means of the Llama, Amaru,

[6] Cerrón-Palomino thinks that "Mama" in this case refers to size rather than source of regeneration (p.c. 2022).

and Machacuay (another serpent) constellations that drink up the excess water each year (Urton 1981a, b). Urton (2021) argues that the celestial axis of the Mayu is primary and foremost in their cosmology. Even the smallest villages were familiar with the motions of the Milky Way in the night skies and were oriented by them. It also provided basic orientation in the universe in time and space for the people who could watch and "read" the great River Mayu, flowing through their skies every night. The circulation of water from the sky to the ocean covering millions of patches of farmed land and pastures depends on this celestial movement. The thunder deity was essential because he could release water from the Mayu so that it would fall as rain.

Other waters flowed down to the Mamacocha (the Atlantic/Pacific Ocean) in the form of surface rivers and subterranean rivers. People could throw away illnesses into the rivers that carried them away to the edge of the earth and return health to themselves with the water from the springs, bogs, and lakes on earth. Lake Titicaca as well as other lakes provide a siphon of water from the ocean up to the earth's surface. These lakes provided the ancestors of people and the original llamas (Sherbondy 1982a, b). Mountains stored water. It was well known that their capacity for storing water and releasing it in springs was a powerful gift of these apu (lords). Surface waters came from the melt of glaciers and snow on high mountain tops. Rainbows stretched from one spring to another to suck up water.

The universe was constantly circulating water, under the earth, inside the earth, on the surface of the earth, up throughout the skies, and down again. People and animals derived life from this unified system. They knew the skies better than we can imagine as well as the interior of the earth. Every water source was revered as a wak'a.

2.6 Conclusions

The importance of water in the world view of the Incas was formulated primarily from the cosmovision of the Middle Horizon Tiahuanaco empire. With the great lake of Titicaca at its center, it was a vision that this world was the result of an Unu Pachacuti, a time of flood that changed the entire society (Bouysse-Cassagne). The great deity Wiracocha drew the new universe out of the womb-like lake, ordering the sun to shine in the sky, apparently after a traumatic time of flooding. The first peoples were also drawn out of their watery origins and Wiracocha sent them by underground stream or channels to emerge at their own special locations on earth (springs, other smaller lakes, caves, mallqui trees) to claim their own territory. This myth probably originated in the Titicaca basin and was the foundation of the religion that was spread by Tiahuanaco. The Incas adopted this framework and placed themselves within it as children of the sun and superiors to other humans.

The Inca official history claims they emerged at Paccaritambo and walked to Cuzco where they settled on a spot in the Huatanay valley where they had command over the best lands and waters. One longer version of this narrative places them in the still

Puquina-speaking Collao kingdom after the Titicaca basin had experienced invasions of Aymara chiefdoms who introduced the Aymara language that had originated in the Wari empire west of Cuzco. The original eight Inca brothers and sisters spoke Puquina as well as Aymara. They experienced a severe drought in the Collao that led to competition with stronger polities that controlled the best water sources. As a result, they emigrated looking for a place with reliable waters and good soils, especially at an altitude favorable for maize. Paccaritambo was not that destination, but Cuzco fulfilled the requirements. They conquered the Rainbow deity of the Cuzco valley and proceeded to establish their settlement between the two major sources of the river. There they built irrigation canals on the higher lands and drainage systems in the marshy river bottom and defeated all the other small polities that had already been settled there. This led to a policy of expansion that was still operating at the time of the Spanish conquest of Cuzco in 1532.

They retained in their religion the deities of the Titicaca basin: the enormous lake as the origin of the universe, the deity Contiti Wiracocha, who was a sun god that became closely associated with the lake, Tunupa/Illapa, thunder and lightning as the bringer of rain, and Inti their own Inca sun. Judging from the Puquina language-based names for deities, they also retained the worship of Huanacauri, the rainbow, and of all the thousands of sacred wak'as, many of them springs, ponds, bogs, and other places associated with water. The worship of the sun was also inherited from Tiahuanaco which had Contiti, the burning hot sun as their solar deity.

This was a natural universe full of water sources and Inca focused religion and rituals were centered on managing the timing and amount of precipitation according to one rainy season and a long dry season, aimed primarily at the cultivation of maize that was a strategic asset of the empire. They organized the whole metropolitan/agricultural area of the valley with wak'as designating sacred places that needed offerings. At least a third of them were overtly water sources. The ayllus and panacas were assigned their social and political rank according to the wak'as they were assigned to care for, thus making the political organization of the entire drainage basin subject to the Inca rulers, the ceque system.

Their knowledge of subterranean waters was extensive and not fully studied as climate, soils and river basins, but it was of great importance to them, not only because it explained origins and migrations but as the hydraulic heart of the earth that can continue to produce water where none appears in the form of rain, either in desert environments or in times of drought. This was their backup reservoir. As people today anxiously observe the melting of the glaciers and snowpacks on the mountain peaks, and late rains, they still may save their crops, herds, and lives with the underground lakes and streams located in and under the mountains. After all, the mountains are worshiped primarily for their water.

Finally, the motor for circulating water was embodied in the celestial Mayu that sucked up excess water from earth, let rain fall to earth, and flowed through the cosmic ocean for replenishment. Its heavenly path was a mirror image of the earthly path of Wiracocha along the fault lines formed by the Nazca plate that created Lake Titicaca and the Huatanay River (Bouysse-Cassagne 1987; Urton 2021; Wachtel 1990).

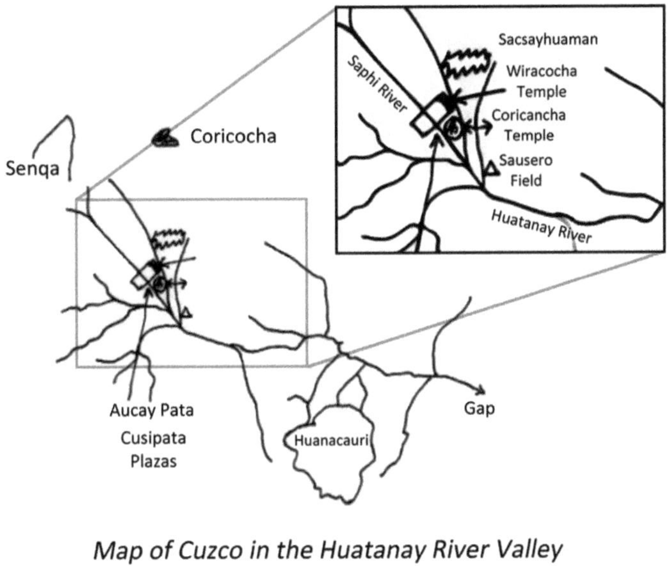

Fig. 2.1 Map of Cuzco in the Huatanay River Valley (illustration by Julia Schreppler)

The Inca legacy of a state organized for agricultural and animal production is one of organizing, observing, respecting, and valuing above all, their water sources and the lands watered by them. Andean villagers are still doing what they can to protect and conserve water, as in the current invention of "breeding and harvesting" of water Apaza Ticona et al. (2021) because "water is life."

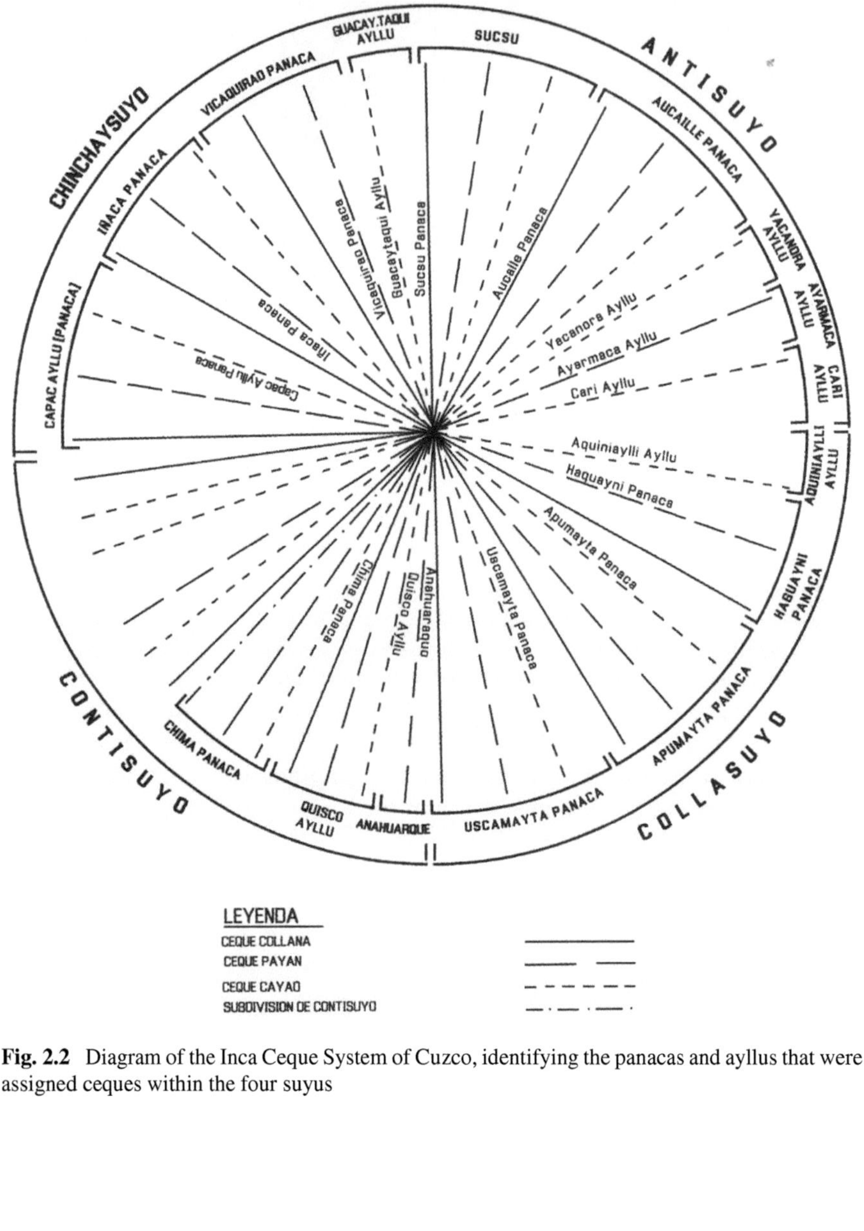

Fig. 2.2 Diagram of the Inca Ceque System of Cuzco, identifying the panacas and ayllus that were assigned ceques within the four suyus

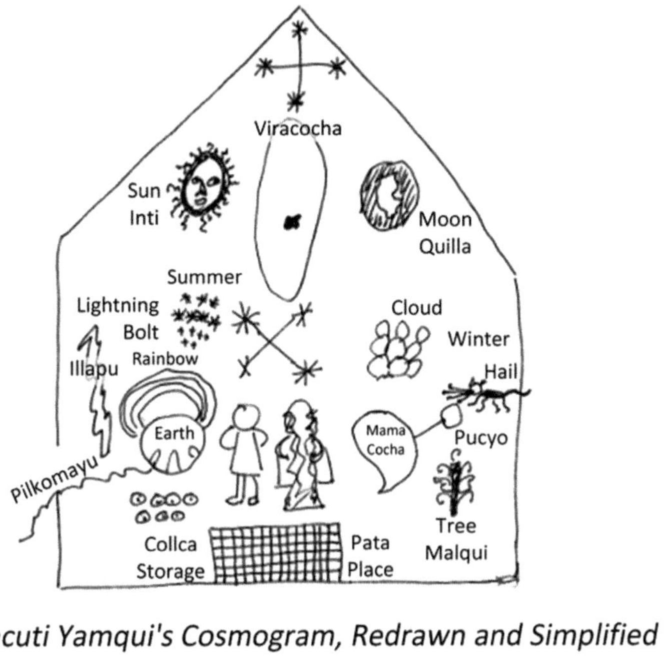

Fig. 2.3 Pachacuti Yamqui's cosmogram, redrawn and simplified (illustration by Julia Schreppler)

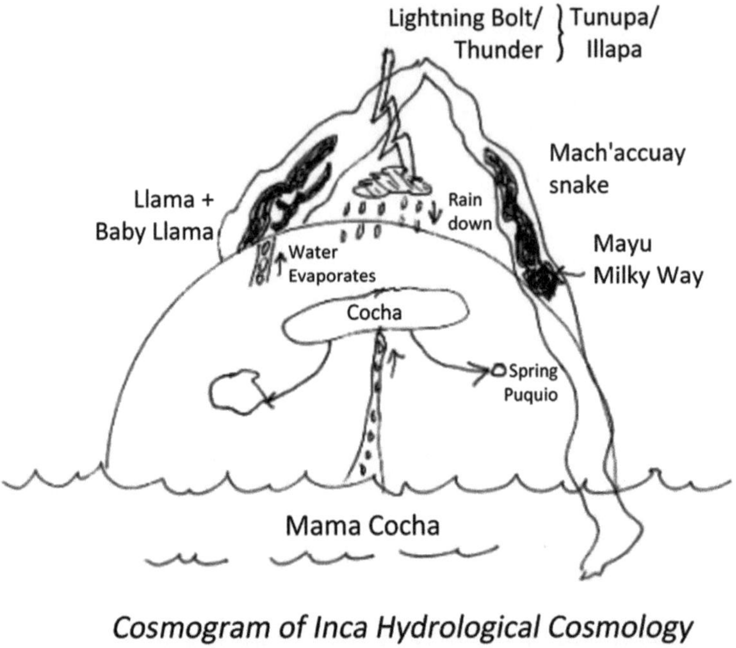

Fig. 2.4 Cosmogram of Inca hydrological cosmology (illustration by Julia Schreppler)

References

Albó J (1972) Dinámica en la estructura intercomunitaria de Jesús de Machaca. América Indígena 32:773–816

Apaza Ticona J, Alanoca Arocutipa A, Calderón Torres G, Cutipa Añamuro (2021) Breeding and harvesting water in the Aymaras communities of Puno-Peru. Academia Lett Article 1689. https://doi.org/10.20935/AL1689

Aveni AF (2000) Between the lines. University of Texas Press

Barreda Murillo L (1973) Las culturas inka y pre-inka de Cuzco. Thesis in archaeology, Universidad Nacional San Antonio Abad del Cuzco

Berenguer Rodríguez J (2000) Tiwanaku: lords of the Sacred Lake. Museo Chileno de Arte Precolombino, Banco Santiago

Betanzos J (1996[1551]) Narrative of the Incas. University of Texas Press, Austin

Bouysse Cassagne T (1987) La identidad aymara: Aproximación histórica (Siglo XV, siglo XVI). HISBOL, IFEA, La Paz

Bouysse Cassagne T (1993) El rayo que no cesa. In: Duviols R (ed) D'un dieu pukina 'a un dieu inca, pp 165–180

Cerrón-Palomiino R (2021) Otra vuelta de tuerca: la etimoloía de <quechua> en quechua y aimara. Indiana 38(2):159–177

Cerrón-Palomino R (2016) El lenguage como hermenéutica en la comprensión del pasado: a propósito del puquina en la génesis del imperio incaico. Diálogo Andino 49:11–27

Cerrón-Palomino R [CP] (2013) Las lenguas de los incas: el puquina, el aimara y el quechua. Peter Lang Academic Research

Cerrón-Palomino R (2018) El puquina como lengua de Tiahuanaco. En MUÑOZ, María de los Ángeles (ed): Interpretando huellas. Arqueología, etnohistoria y etnografía de los Andes y sus

tierras bajas. Cochabamba: Universidad Mayor de San Simón/ Fundación Simón I. Patiño/ Grupo Editorial Kipus, pp 415–428
Cerrón-Palomino R (2020a) La presencia puquina en el Formativo Tardío en el valle del Cuzco. In: Vega Centena RY, Dulanto, J (eds) El tiempo, el espacio y la memoria. Ensayos en homenaje a Peter Kaulicke. Lima, Fondo Editorial PUCP, pp 71–89
Cerrón-Palomino R (2020b) Los nombres primigenios del Cuzco. Revista Histórica, XLIX, 113–132
Cerrón-Palomino R (2021b) Los riesgos de una lingüística desmemoriada: a propósito de la etimología puquina de <inca>. Lexis XLV(1):227–261
Cieza de León P (1959[1553]) The Incas. University of Oklahoma Press
Cobo B (1990[1653]) Inca religion and customs. University of Texas Press
Domínguez Faura N (2017) Prefacio. In Sherbondy 2017:14–25
Duviols P (1993) Religions des Andes et langues indig'enes: Équateur-Pérou-Bolivie, avant et après la conqu^ete espagnole. L'Université de Provence
Duviols P (2008) Pachacuti Yamqui Salcamaygua, Joan de Santa Cruz (seventeeth century). In: Pillsbury J (ed) Guide to documentary sources for Andean studies, 1530–1900, vol 3, pp 488–496
Earls J, Silverblatt I (1976) La realidad física y social en la cosmología. Actes du XLII Congress International des Americanistes. Paris, Septembre 1976
Glowacki M (2012) Imperialismo en el horizonte medio: Una reevaluación del paradigma clásico, Cuzco, Perú. Boletín De Arqueología PUCP 16:189–207
Janusek JW (2004) Tiwanaku and its precursors: recent research and emerging perspectives. J Archaeol Res 12(2)
McEwan GF (2005) Pikillacta: the Wari empire in Cuzco. University of Iowa Press
Molina C (1943[1573]) Fábulas y ritos de los Incas. n.p.
Murra J (1980) The economic organization of the Inka state. Jai Press
Plourde AM, Stanish C (2005) The emergence of complex society in the Titicaca basin: the view from the North
Polo de Ondegardo J (1917 [1571]) Informaciones acerca de la religión y gobierno de los Incas. Colección de Libros y Documentos Referentes a la Historia del Perú, vol 4
Guaman Poma de Ayala [GP] (1980 [1615]). El Primer nueva corónica y buen gobierno. Siglo Veintiuno, Instituto de Estudios Peruanos
Pachacuti Yamqui Salcamaygua J [PYS] (1993[1613]). Relación de antigüedades deste reyno del Piru. IFEA y CERA "Bartolomé de Las Casas", Cuzco
Sherbondy JE (1986b) Los ceque: Código de canales en el Cusco incaico. Allpanchis 27:39–74
Sherbondy JE (1992) Water ideology in Inca ethnogenesis. In: Dover R, Seibold K, McDowell J (eds) Andean cosmologies through time: persistence and emergence. Indiana University Press, Bloomington, pp 46–66
Sherbondy JE, Villanueva UH (1979a) Cuzco: Aguas y poder. Archivos de Historia Rural Andina, vol 1. Centro de Estudios Rurales Andinos Bartolomé de Las Casas, Cuzco
Sherbondy JE (1979b) Les réseaux d'irrigation dans la geographie politique de Cuzco. Journal de la Societé des Américanists 66:45–66. Paris, France
Sherbondy JE (1982a) The Canals of Hanan Cuzco. Ph.D. dissertation, University of Illinois at Champaign-Urbana
Sherbondy JE (1982b) El regadío, los lagos y los mitos de origen. In: Religión, mito y ritual en el Peru, special issue of Allpanchis 20:3–32. Cuzco, Peru: Instituto Pastoral Andino
Sherbondy JE (1986a) Los ceque: Código de canales en el Cusco incaico. Allpanchis 27:39–74
Sherbondy JE (1986b) Mallki: Ancestros y cultivo de árboles en los Andes. Apoyo a las plantaciones forestales con fines energéticos y para el desarrollo de comunidades rurales de la Sierra Peruana, Proyecto FAO- Holanda/lNFOR GCP/PER/027/NET Documento de trabajo 5, Lima, Peru
Sherbondy JE (1987) Organización hidráulica y poder en el Cuzco de los incas. Revista Española de Antropología Americana, vol 17. Complutense University, Madrid, Spain, , pp 117–153
Sherbondy JE (1987b) The Incaic organization of terraced irrigation in Cuzco, Peru. Br Archaeol Rep Int Ser 359(1):365–371. In: Proceedings of the 45th international congress of Americanists, Bogotá, Colombia 1985, Pre-Hispanic Agricultural Fields in the Andean Region, Part 1

Sherbondy JE (1988) Ancestros y cultivos de árboles en los Andes. Sociedad andina pasado y presente, vol 101–135. FOMCIENCIAS, Lima
Sherbondy JE (1994) Water and power: the role of irrigation districts in the transition from Inca to Spanish Cuzco. In: Wm P, Mitchell WP, Guillet D (eds) Irrigation at high altitudes: the social organization of water control systems in the Andes. American Anthropological Association, pp 69–97
Sherbondy JE (1995) El agua: Ideología y poder de los Incas. In: González Alcantud JA, Malpica Cuello A (eds) El agua. Mitos. ritos y realidades. Barcelona: Anthropos, pp 87–102
Sherbondy JE (1996) Panaca Lands: re-invented communities. Structure, knowledge and representation in the Andes: studies presented to Reiner tom Zuidema on the Occasion of His 70th Birthday, theme issue. J Steward Anthropol Soc 24(1 and 2):173–202
Sherbondy JE (1998) Andean irrigation in history. In: Jose Almeida J, Boelens R (eds) Searching for equity: peasant conceptions of equity and justice in peasant irrigation. Van Gorcum SNV-Ecuador, pp 210–214
Sherbondy JE (2017) Agua Riego y Árboles: Ancestros y Poder en el Cuzco de los Incas. Sociedad Geográfica de Lima
Silverblatt I (1987) Moon, Sun, and Witches: gender ideologies and Class in Inca and Colonial Peru. Princeton University Press
Urton G (1981b) Animals and astronomy in the Quechua universe. Proc Am Philos Soc 125(2):110–127
Urton G. (1981a) At the crossroads of the earth and the sky: an Andean cosmology. University of Texas Press
Urton G (1990) The history of a myth: Pacaritiqtambo and the origin of the Inkas. University of Texas Press
Urton G (2021) Andean Cosmos. Academia.edu
Valencia Zegarra A (2005) Wari hydraulic works in the Lucre Basin. In: McEwan G
Wachtel N (1990) Le retour des anc^etres. Éditions Gallimard
Zuidema RT (1990) Inca civilization in Cuzco. University of Texas Press
Zuidema RT, Urton G (1978) La constelación de la llama en los Andes peruanos. Allpanchis 9:59–119
Zuidema RT (1964) The ceque system of Cuzco. E.J. Brill, Leiden
Zuidema RT (2010) El calendario inca: Tiempo y espacio en la organización ritual del Cuzco. La idea del Pasado. Fondo Editorial del Congreso del Perú, Fondo Editorial de la PUCP

Chapter 3
Mayans and Their Water Management

N. A. Oropeza-García, R. E. Vega-Azamar, B. L. Campos-Cámara, J. A. Canché-Uuh, and M. A. González-Vera

Abstract The Mayan culture was established in the jungle of the Yucatan Peninsula in Mexico, extending to Guatemala, Belize, and part of Honduras and El Salvador. In the Mayan worldview, water appears as a central forming element of the Universe and as the Earth's soul, being a sacred life-giving element. God *Chaac*, creator of lightning, rain and water, carried an axe with which he broke jugs releasing water to make it fall. According to Mayan stories, *Chaac* inhabited caves and sinkholes (known locally as *cenotes*, from the Mayan word *ts'ono'ot*), places that were considered the entrance to the underworld or world of the dead. Due to the karst soil, widely predominant in the Yucatan Peninsula, the region does not have surface runoffs, such as rivers or lakes, being the underground flows and *cenotes* (considered sacred places for the birth of life because they contained virgin water) the main sources of water supply. In addition to *cenotes*, in relation to water management among the Mayans, evidence has been found of the development of hydraulic systems to face scarcity in the dry season, which included canals, dams, and irregular cisterns called *chulub tun*, where they caught and stored rainwater for later use.

Keywords Artificial reservoirs · Karst soil · *Cenote* (sinkhole) · Nature-based ancient hydraulic systems

3.1 Introduction

The Mayans inhabited what is now southeastern Mexico, covering the coastal plains of the state of Tabasco, the highlands of the state of Chiapas, the lowlands of the state of Campeche, the Yucatan Peninsula, and parts of Guatemala (except certain lowlands on the Pacific coast), Belize, the western border of Honduras, and the

N. A. Oropeza-García · B. L. Campos-Cámara · J. A. Canché-Uuh · M. A. González-Vera
Universidad Autónoma del Estado de Quintana Roo, Chetumal, Mexico
e-mail: noropeza@uqroo.edu.mx

R. E. Vega-Azamar (✉)
Instituto Tecnológico de Chetumal / Tecnológico Nacional de México, Chetumal, Mexico
e-mail: ricardo.va@chetumal.tecnm.mx

G. A. Corzo Perez et al. (eds.), *Aqua Vitae*, The Latin American Studies Book Series,
https://doi.org/10.1007/978-3-031-57731-4_3

northeastern part of El Salvador. In general, this vast area can be subdivided into highlands and lowlands (Fig. 3.1).

The highlands occupied by the Mayans are located in the territory of Guatemala and the Mexican state of Chiapas, in the mountain systems of *Sierra de los Cuchumatanes* and *Sierra Madre* reaching heights of up to 3800 m.a.s.l., presenting a temperate climate, coniferous vegetation and sources of streams and rivers in their fertile valleys, which were used for agriculture. In these places, the mountains themselves were defined as religious sites, so large temples are not observed. In the highlands, the Mayans exploited mineral deposits located in river basins, ravines and volcanic outcrops, among others. An important site in this area for them was the

Fig. 3.1 Territory occupied by the Mayan culture and main settlements (Wells 2016) (Coordinates not provided)

Motagua River Valley, which contains the only jade deposits in the entire American continent. The archeological remains in this area are not considered very representative of the Mayan development (250–950 AD) due to the cultural influence of the Mexican High Plateau (Terán and Rasmussen 2009).

For their part, the lowlands are found in the Yucatan Peninsula, the alluvial plains of Tabasco, the north of Chiapas and the west of the Usumacinta River in Mexico, and extending to Guatemala, Honduras and Belize. In this area, the altitude does not exceed 300 m.a.s.l., presenting a warm-humid climate and tropical vegetation. This is an area of great archeological interest because the largest and most complex human settlements were developed there, showing elements of an elaborated culture with its own and refined expressions in cities such as *Tikal*, *Copan*, *Bonampak*, *Palenque*, *Piedras Negras* (150–1200 AD), among others (Terán and Rasmussen 2009).

3.2 The Mayan Culture

In pre-Columbian America, the Mayan civilization is one of the most advanced and complex; from the intellectual and artistic point of view, it was the only one that dominated true writing, considered the most complex in all America; they calculated time from an imaginary point of origin. The history of the Mayan civilization has been divided into three major periods of time: Preclassic (2000 BC–100 AD), classic (250–950 AD), and post-classic (950–1525 AD), leaving important cultural contributions that survive to this day in the sites they inhabited (Baudez 2004).

The Mayans developed their culture from the natural resources surrounding them and generated amazing knowledge for the use of vegetation, fauna, minerals and celestial bodies (SARH-IMTA 1988). The close relationship of the Mayans with nature was the basis of their religious thoughts and the Mayan Worldview, which is reflected in their deities associated with different elements of nature such as: *Chaac* the god of rain, *Yum-Kax* the god of corn, *Kukulcan* the god of the wind, *Ah Puch* the god of the jungle, *Xaman Ek* the god of the Polar Star, *Yum Cimil* the god of death, *Kinich Kakmo* the god of the sun, *Ixchel* the goddess of the Moon.

Water sources have always been a benchmark for human settlements. In the case of the Yucatan Peninsula, the Mayans settled near large *cenotes* or floodplains, which they identified as *a'kalche*. In the Mayan culture, no large constructions have been observed by the sea, with the exception of Tulum, which is built on a massive cliff of calcareous rock. The Maya built their settlements inland taking advantage of the resources of the coastal dune and wetlands by building temporary camps and thus avoiding the impact of hurricanes (Correa-Sandoval 2011). Currently, wetlands are considered one of the most productive ecosystems in the world, they are a source of freshwater, and they provide ecosystem services such as groundwater recharge, hurricane buffering, carbon dioxide absorption, among others. The definition of wetland includes lakes and rivers, underground aquifers, swamps and marshes, wet grasslands, peatlands, oases, estuaries, deltas and tidal flats, mangroves and other coastal areas, coral reefs, and human-made sites such as fish ponds, rice paddies, reservoirs, and salinas (RAMSAR 2011).

In the pre-Hispanic Mayan culture, three main sources for water supply have been documented (Chávez-Gómez and Icaza-Lomelí 2010):

Natural sources: rivers, lagoons, *petenes*, ponds, *cenotes*, *rejolladas*, *sartenejas* or *haltunes*, *akalches*, and *xuayabas*.
Catchment infrastructure: *chultunes*, wells, dams, and canals.
Modified natural reservoirs: bordered waterways, built accesses, and stucco coatings.

One of the ceremonies that are preserved to this day is that of *Cha-Chaac*, which is performed to ask the god *Chaac* for rain, it is very important for the Mayan peasants. In the Mayan culture, the man forms a whole with the cosmos and does not assume himself as the owner or tamer of nature to exploit it for his own benefit, but instead worships it, asking for its permission and help, thanking the goods granted by the Mother Land (Morales-Damián 2010).

3.3 Water in the Yucatan Peninsula

The Yucatan Peninsula does not have mountains; it is made up of large plains and some slopes and elevations not exceeding 500 m.a.s.l. Due to its predominantly limestone stony soil type, with an important presence of slabs, it is estimated that around 80% of the rainwater infiltrates the subsoil forming deposits and underground rivers; rainfall ranges between 800 and 1200 mm per year (Ríos-Ponce et al. 2020). The only river system in the Yucatan Peninsula is Hondo River, which is located on the border between Mexico and Belize. Hondo River is born in Guatemala with the name of Blue River and is actually a geological fault, 85% of its flow is of underground origin and the remaining 15% is attributed to rainfall (Álvarez-Legorreta 2011).

The most important feature in the topography of the Yucatan Peninsula is the presence of *cenotes*. The highest concentration of *cenotes* is found in the northern part along an imaginary line that goes from Tulum in the Mexican state of Quintana Roo to the state of Campeche which is located in the southern part of the peninsula. The word *cenote*, meaning cavern with a water reservoir, comes from the Mayan *ts'ono'ot* or *d'zonot* and is associated with any space in which underground water can be observed and in some of its spots is open to the outside. Due to the scarce relief of the Yucatan Peninsula and the high permeability of the karst soil, the hydraulic gradient in the soil is practically zero and rainwater infiltrates rapidly, forming a layer of fresh water that floats on a mass of saline water of greater density coming from the Caribbean Sea and the Gulf of Mexico which, by intrusion, reaches the continental zone of the peninsula. The fresh water layer that forms in the limestone soil is the only source of freshwater in the Yucatan Peninsula and its renewal, therefore, depends on the rainy season throughout the year. Due to the type of limestone rock and the scarcity of soil of organic origin in the peninsula, the aquifer is very vulnerable to contamination (Schmitter-Sotto et al. 2002).

In the Yucatan Peninsula, water circulates through fractures and underground channels moving from zones with greater precipitation, in the center-south of the peninsula, flowing toward the coast, recharging underground aquifers, wetlands and coastal lagoons as it passes (Álvarez-Legorreta 2011).

The process leading to the formation of *cenotes* is called karst, which consists of the combination of several mechanisms of dissolution, collapse, and reconstruction of the limestone rock advancing from the surface, due to the water infiltration in a horizontal system. Formation of *cenotes* is a dynamic and complex chemical corrosion process in which soluble rocks (gypsum, limestone, dolomite, and halite) are diluted, resulting in the formation of different types of *cenotes* depending on the climatic conditions (temperature, fresh and salt water concentrations, and duration) and soil characteristics (rock porosity, bedrock fractures, vegetation type) (Schmitter-Sotto et al. 2002).

Cenotes are classified according to the stage of the opening process. That communicates the underground aquifer with the ground surface and to the access to sunlight. Figure 3.2 presents a classification of the types of *cenotes* according to their geological evolution: (1) Anquihaline cave, (2) lotic jug *cenote* type or young *cenote*, (3) vertical wall *cenote*, (4) lentic or old type *cenote*, (5) *cenote* with sediment and sinking of walls, and (6) pond-type *cenote* (Schmitter-Sotto et al. 2002).

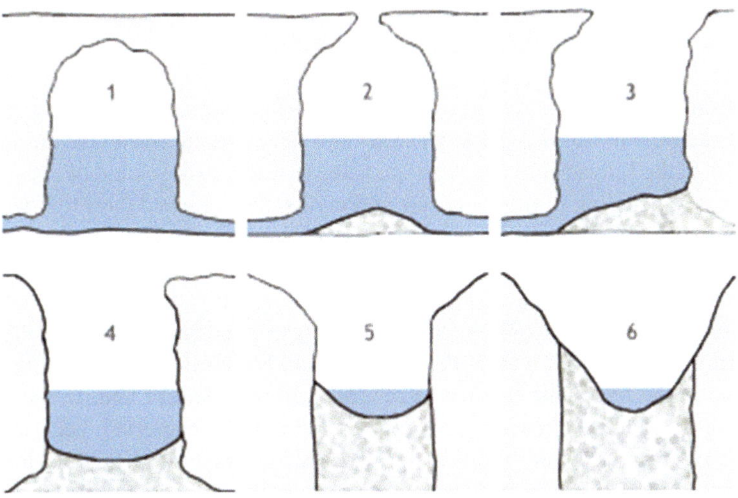

Fig. 3.2 Classification of cenotes, according to the stage of opening (Schmitter-Sotto et al. 2002)

3.4 Mayan Worldview and Water

Pre-Hispanic cultures in Mesoamerica, like others in the world, are based on their growth and development on the use of natural resources. In the case of water, it came from different sources: rainwater, surface water from rivers, springs, streams, wetlands, lagoons, swamps, and groundwater. In this way, in the Yucatan Peninsula, as well as in other lowland areas, *cenotes* were the key to the establishment of villages and cities (IMTA-CIESAS 2009).

In the vision of Mayan cosmology, water must be kept clean and must not be paid for because it is part of nature and a blessing from their gods. Water is an entity with its own will and magical beings inhabit it. In the Mayan cities, the ponds and water bodies provided a projection to the "other world," through their reflection, for temples, palaces, and ceremonies, thus materializing their rituals in that magical and spiritual world associated with the underworld, connecting with the world of the dead and maintaining contact between the underworld and the world of men (Rivera-Dorado 2001).

In the Mayan cities, *cenotes* were part of the sacred places where rituals, offerings and sacrifices were performed to ask for rain to *Chaac*, who arose from the *cenotes*. In Chichen Itza (or well's mouth of the water wizards), the *cenote* of this important settlement, is an example of the adjustments made by the Mayan engineers and architects to modify *cenotes*, turning it into a place for the public to participate in the ceremonies, in this way, *cenotes* received improvements and modifications for their conservation, given the valuable source of water that they represented and because they were considered sacred places (Piña-Chan 2013).

Water from *cenotes* is purer and more suitable for drinking, unlike water from wells, from the *Áak'al* or *aguadas* (ponds) and from *haltunes* or *sartenejas* (holes formed naturally in stones or rocks) where rainwater is stored. Due to the difficulty in obtaining it, the Mayan culture gave great value to water, to the extent of considering it divine; by including it in their rituals and myths, they allowed their link with this element to be strengthened (Duarte-Duarte 2015).

Cenotes are the entrance to the underworld, whoever enters the earth through a *cenote* comes into contact with tectonic forces, germination and with the gods of death, and, according to the Mayan cosmology, enters the underworld of death to return to the maternal womb, the womb of Mother Earth, surrounded by water, returning to the fetal state to later be reborn from the Sacred Land. In the Mayan culture of Yucatan, *cenotes* are sacred places that must be respected, because there, the gods keep sacred water, from which *Yum Chaac* draws it and then drops it as rain. Among the magical beings that inhabit the *cenotes*, are the *chaaces* or rain deities, the *aluxes* or jungle elves and *Tsukan* or giant snake in charge of guarding the aquifer (Rojas-Sandoval 2013).

3.5 Water Management in the Mayan Culture

Mayans classified water depending on its source: seawater, rainwater, and surface bodies; in the Mayan language: *Kaknab*, salty water from the sea; *Chulub*, rainwater and *Ha*, water from surface currents different from water from *cenotes* and wells, and according to characteristics such as flavor and color: sweet or salty, and black, green or transparent (Chávez-Gómez and Icaza-Lomelí 2010). Water management by the Mayans was focused on meeting their needs: supply for domestic use, conduction and drainage to avoid flooding and contamination in their settlements, supply for irrigation of their cultivated fields, and provision for their religious ceremonies. Therefore, ancestral societies developed knowledge about the temporality of rains, the quality of water depending on the source, and how to optimize its use according to their needs.

The main process on which the Mayan culture based its water management was the collection and storage of rainwater during the rainy season. Water needed for drinking, cooking, building and performing ceremonial rituals, as well as healing, was an essential part of this culture. The Mayan people based their organization around dedication and worship to water. Mayan priests and architects designed public service infrastructure that allowed them to collect rainwater to overcome the crises and shortages suffered from generation after generation. However, to ensure that each year water resources for crops and livelihood of the population were guaranteed, the rulers had to perform ceremonies and make offerings to the god *Chaac*. A city that shows ingenious constructions to collect rainwater is Uxmal, located in the northern part of the Yucatan Peninsula. The "Quadrangle of the Nuns" may be observed as part of the structures for the collection and storage of rainwater in Uxmal; also, numerous *chultunes* can be found everywhere (Romero-Infante 2006).

3.5.1 *Chultun*

Chultunes are underground deposits built by the Mayans, from limestone rock carving, to store rainwater, which allowed them to keep it fresh and purer for domestic use in times of drought and scarcity (Fig. 3.3). Thousands of *chultunes* built on the Yucatan Peninsula's karst soils have been documented. *Chultunes* maintained a catchment area covered with stucco sloping toward the mouth of a cistern with a diameter of approximately 50 cm (IMTA-CIESAS 2009).

Another characteristic, cited in the literature, is the delimitation of a catchment perimeter with aligned stones forming concentric circles of approximately 1.30 cm in diameter and, depending on the area where the chamber is located, covered with stucco made from *sahcab* (sandy limestone) or by courses of carved stone. *Chultunes* could be supplied with water through gutters from the roofs and from surface canals that directed the runoff from a ceremonial area toward the stucco-covered catchment

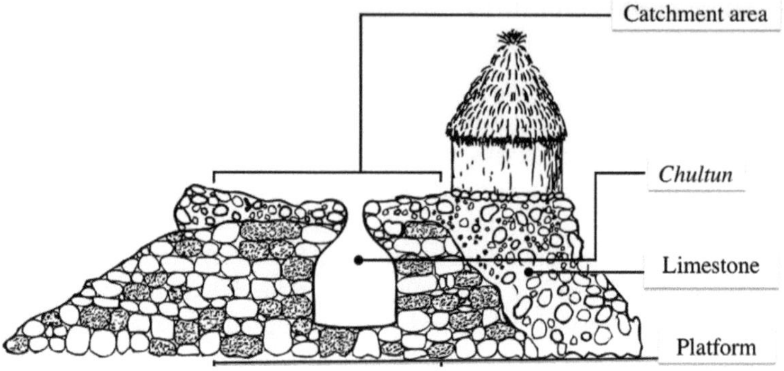

Fig. 3.3 Profile of a chultun (INAH 2019)

area; the storage chambers had different shapes: amorphous, bottle, vault and bell (Zapata 1989).

Chultunes have been found in different Mayan cities such as Uxmal, Nakbe, Edzna, Kabah, Chacmultun; they are well preserved to this day. *Chultunes* were an important urban component in Mayan cities because they provided their inhabitants with the vital liquid, putting it at the service of the community and also possessed a symbolism of abundance associated with the connection between the gods and a magical and ritual space. *Chultunes* have been located mainly in areas where the presence of *cenotes* is not observed, in the *Peten*, the *Chenes*, and the *Puc* Zone. *Chultunes* were also used to store food such as corn and to ferment alcoholic beverages, and when their lifespan ended, the Mayans used them to bury their dead (Rivera-Dorado 2001).

3.5.2 Haltun

Sarteneja or *haltun* (water in stone) is a hole in the limestone rock, formed naturally by the dissolution of the rock, in which water was collected during the rainy season for use by peasants and domestic animals (Fig. 3.4). Usually, in this kind of formation, there is no contact with the water table; however, depending on the size and depth, it can contain water during the dry season (Flores-Guido 2010). An example of this type of receptacle, formed naturally or that may also have been sculpted by the ancient Mayans, was documented in the Calcetoh caves in the Mexican state of Yucatán to collect runoff (Méndez-Granados 1999). *Haltunes* were classified by the Mayans depending on their shape, depth and extension, and the water collected had different purposes:

Chen haltun: hollow formed in the rock to catch water, circular in shape with a style similar to a deep bowl, which was used for human supply.

Fig. 3.4 Haltun or sarteneja (Explore Yucatan 2021)

Nohaltun: shallow oval-shaped hole present in the limestone rock where water was accumulated for animal consumption or for agricultural purposes due to the presence of sludge.

Tsono'haltun: wide and shallow pool used to capture water for crops and animals due to the presence of sediments.

3.5.3 Other Sources of Water Supply

In places like the city of Chichen Itza, there are three sources for water supply (IMTA-CIESAS 2009).

Rejolladas (Fig. 3.5) are cavities similar to *cenotes* that may have originated from them when the water mirror was lost or due to the dissolution of carbonates (Flores-Guido 2010).

Aguadas (ponds) are formed in shallow lands where rainwater is deposited, the Mayans adapted them with stucco coatings and by building borders and accesses, so it was necessary to clean and clear them before each rainy season. *Aguadas* can also be man-made with a depth of 5–8 m.

Wells are perforations made in the ground until reaching underground water, sometimes reaching a depth of up to 20 m (Chávez-Gómez and Icaza-Lomelí 2010). Peninsular Mayans drilled holes in the calcareous rock until they reached the water table, a very valuable representation of water use combining different alternatives (Fig. 3.6) shows an *aguada* in which wells and *chultunes* were included, which allowed, on the one hand, to catch rainwater in the rainy season and to capture water from the groundwater table by infiltration during the dry season (IMTA-CIESAS 2009).

Fig. 3.5 Rejollada and associated vegetation (Flores-Guido 2010)

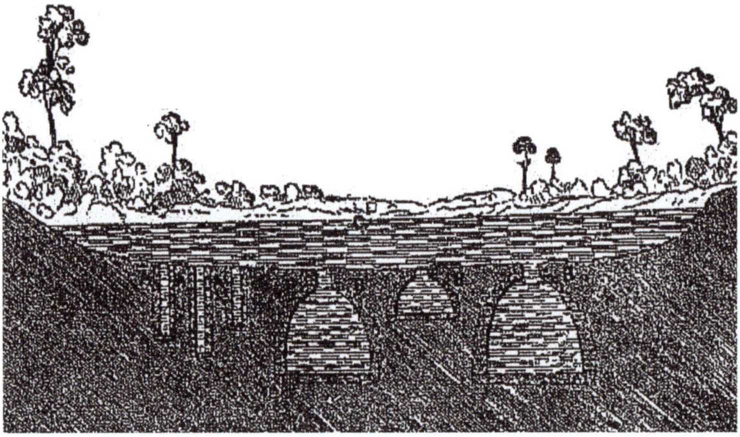

Fig. 3.6 Combined water reservoirs: aguada, well and chultun (IMTA-CIESAS 2009)

3.5.4 Hydraulic Systems in Mayan Cities

Canals, drains, reservoirs, aqueducts, and slopes are part of the hydraulic system observed in ancient Mayan cities such as Edzna, Palenque, Joy Chan (Comalcalco), Chichen Itza, and Palenque (French 2007), which show the construction of their important buildings on platforms conceived to create slopes that channeled water to drains to avoid the flooding of squares and esplanades of the cities. The water that was drained from the cities was led to low areas modified to capture and retain it in *rejolladas* or sinkholes. An important adaptation to capture rainwater and prevent infiltration can be seen on the roofs of buildings that have a slope and stone gutters in the moldings that channeled the rainwater; all these characteristics are common in both cities of lowlands and highlands of the Mayan region as well as in other areas of Mesoamerica (González de la Mata et al. 2005).

3.5.5 Water and Agriculture

The use of water in agriculture was applied to different microenvironments. Among the archeological remains, three cultivation methods have been documented in the south of the Mexican states of Campeche and Quintana Roo, which include (Turner 1981).

Irrigation networks, through the construction of platforms and canals of soil and weathered rock that prevented flooding and irrigated crops in the dry season. Cities showing evidence of irrigation and drainage for domestic uses are Edzna, Tikal, and Copan.

Raised fields in riparian areas and temporarily flooded depressions led to the creation of orchards and raised fields to drain the flooded lands and maintain a moisture regime that intensified food production. Vestiges of this type of cultivation are documented in the northeast of Peten, toward Quintana Roo and Belize.

Linear hillside terraces, some without irrigation, and channeled bottom terraces. The use of terraces allowed the ancient Mayans to establish the thickness of the soil and increase humidity by keeping the water in the drains for longer periods of time. Vestiges of this type of cultivation technique are particularly mentioned for those used in the Bec River area, in the Mexican state of Campeche, to prevent erosion in lands with little inclination.

These agricultural technologies in the highlands and in the lowlands are basically reported as the same, the most important difference between the two regions being the need for irrigation of crops, which led the Mayan culture to have an intensive agriculture of several harvests a year, a situation associated with sociopolitical growth as a consequence of increased productivity, population and an increase in "energy" in the organization.

3.6 Conclusions

The Mayan culture was one of the most fruitful of its time. The Mayans made mathematical, astronomical, and medical contributions from their use and observation of vegetation, fauna, minerals and celestial bodies. Their water management was focused on supply for domestic use and irrigation, conduction and drainage to prevent flooding and contamination, and religious purposes. Mayans developed knowledge about temporality of rains, water quality based on the available sources, and use optimization. They observed the organoleptic characteristics of the water, classifying it into sea water, green water, black water and transparent water, defining uses for irrigation, animal or human consumption.

Through observation and tracking of natural phenomena, the Mayans avoided settling in coastal zones (with the exception of Tulum) and used temporary camps to take advantage of the resources derived from the dune and mangrove, which is a reservoir for different animal species, thus maintaining this ecosystem as a source of usable resources and, at the same time, avoiding the attack of hurricanes by settling inland.

In their water supply systems, it is possible that the ancient Mayans were inspired by the natural formation of *cenotes* for the construction of *chultunes*, which are similar in their jug-shaped interior. An interesting element of *chultunes* is the construction of concentric circles in the catchment area, which would lead to the settling of fine particles when they collide with the inner circle walls, thus obtaining water with a smaller amount of particulate matter and, consequently, of better quality for consumption.

Another phenomenon that they should have noticed was filtration occurring naturally when rainwater passes through the limestone rock, so they defined the *cenotes* water as the purest and most sacred of all, which they used for their rituals and for other medicinal purposes.

In the most important settlements, they planned, from the basement of their buildings, the conception of canals to capture rainwater and prevent flooding in their squares and ceremonial sites, leading it to lower places to form water reservoirs in areas previously covered with stucco or identifying low permeable areas, thus storing the vital liquid for the dry season.

Without a doubt, the Mayan culture is a historical reference in water management and observation of natural phenomena associated with floods, hurricanes, and droughts that should be reviewed thoroughly in the current era in which around 10% of the world population lives in coastal areas where rainwater is poorly used in cities covered with impermeable materials, usually mixing storm drainage systems with sanitary sewage systems, wasting this important resource.

References

Álvarez-Legorreta T (2011) Capítulo 3, Sección: Uso y manejo de los recursos naturales [Chapter 3, Section: Use and management of natural resources]. In: Pozo C, Armijo-Canto N, Calmé S (eds) La riqueza biológica de Quintana Roo [Biological wealth of Quintana Roo] vol 1. CONABIO, ECOSUR, Mexico, p 121. ISBN 978-607-7607-46-5

Baudez CF (2004) Una historia de la religión de los antiguos mayas [A history of the ancient Mayan's religion]. Mexico's National Autonomous University, Institute of Anthropological Research, French Center for Mexican and Central American Studies, 1st edn. UNAM. México, p 427. ISBN 970-32-1575-0

Chávez-Gómez JM, Icaza-Lomelí L (2010) Norias de Yucatán. Ensayo sobre el mestizaje de una síntesis geométrica [Norias of Yucatan. Essay on the miscegenation of a geometric synthesis]. Boletín de Monumentos Históricos [Historical Monuments Bulletin] Issue 18, p 34 INAH [Mexico's National Institute of Anthropology and History], Mexico

Correa-Sandoval J (2011) Capítulo 3, Sección: Humedales [Chapter 3, Section: Wetlands]. In: Pozo C, Armijo-Canto N, Calmé S (eds) La riqueza biológica de Quintana Roo [Biological wealth of Quintana Roo], vol 1. CONABIO, ECOSUR, Mexico, p 121. ISBN 978-607-7607-46-5

Duarte-Duarte A (2015) La "Santa Agua" en la cultura maya en Yucatán [The 'Holy Water' in the Mayan culture of Yucatan]. In: El manejo del agua a través del tiempo en la Península de Yucatán [Water management through time in the Yucatan Peninsula]. Universidad Autónoma de Yucatán, Centro de Investigaciones Regionales Dr. Hideyo Noguchi, Unidad de Ciencias Sociales, Fundación Gonzalo Río Arronte IAP, Consejo de Cuenca de la Península de Yucatán

Flores-Guido S (2010) Capítulo 3. Ecosistemas y comunidades [Chapter 3. Ecosystems and communities]. In: Durán R, Méndez M, Cicy M (eds) Biodiversidad y desarrollo humano en Yucatán [Biodiversity and human development in Yucatan]. PPD-FMAM, CONABIO, SEDUMA. Mexico, p 496

French K (2007) Proyecto Hidro-Arqueológico de Palenque [Palenque Hydro-Archaeological Project]. FAMSI, Foundation for the Advancement of Mesoamerican Studies. Retrieved on November 10, 2021 from: www.famsi.org/reports/05076es/index.html

González de la Mata R, Osorio JF, Schmidt PJ (2005) El flujo divino: Manejo del agua en Chichen Itza [The divine flow: Water management in Chichen Itza]. In: Laporte JP, Arroyo B, Mejía HE (eds) XVIII Simposio de Investigaciones Arqueológicas en Guatemala [XVIII Archaeological Research Symposium in Guatemala], vol 2004. Museo Nacional de Arqueología y Etnología [National Museum of Archeology and Ethnology], Guatemala, pp 847–855

IMTA-CIESAS (2009) Cultura hidráulica y simbolismo mesoamericano del agua en México prehispánico [Hydraulic culture and Mesoamerican symbolism of water in pre-Hispanic Mexico], 1st edn. Instituto Mexicano de Tecnología del Agua [Mexican Institute of Water Technology], Centro de Investigaciones y Estudios Superiores en Antropología Social [Center for Research and Higher Studies in Social Anthropology]. Mexico, p 298. ISBN 978-697-7563-06-8

Méndez-Granados D (1999) Manejo del agua entre los mayas de la península de Yucatán [Water management among the Mayans of the Yucatan Peninsula]. In: El agua en la cosmovisión y terapéutica de los pueblos indígenas de México [Water in the cosmovision and therapy of the indigenous peoples of Mexico]. Instituto Nacional Indigenista [Mexico's National Indigenous Institute], 1st edn. Mexico, p 147. ISBN 978-970-1838-84-6

Morales-Damián MA (2010) Territorio sagrado: cuerpo humano y naturaleza en el pensamiento maya [Sacred territory: human body and nature in Mayan thought]. Cuicuilco 17(48):279–298. Retrieved on October 21, 2021 from: http://www.scielo.org.mx/scielo.php?script=sci_arttext&pid=S0185-16592010000100014&lng=es&tlng=es

Piña-Chan R (2013) Chichén Itzá. La ciudad de los brujos del agua [Chichen Itza. The city of the water wizards]. Fondo de la Cultura Económica [Economic Culture Fund], 2nd edn. Mexico, p 174. ISBN 978-607-1613-31-8

RAMSAR (2011) La importancia de los humedales [The importance of wetlands]. The Ramsar Convention on Wetlands. Retrieved on November 9, 2021 from: https://www.ramsar.org/es/acerca-de/la-importancia-de-los-humedales

Ríos-Ponce JD, Acosta-González G, Cejudo E (2020) La precipitación histórica y la extracción del agua subterránea en la Península de Yucatán: una reflexión [Historical precipitation and groundwater extraction in the Yucatan Peninsula: A reflection]. Desde el Herbario CICY [Yucatan Scientific Research Center] 2020(12):110–118, ISSN 2395-8790

Rivera-Dorado M (2001) La ciudad maya, un escenario sagrado [The Mayan city, a sacred scene], 1st edn. Editorial Complutense, Madrid, p 348. ISBN 978-847-4916-23-2

Rojas-Sandoval C (2013) El inframundo maya y los sistemas kársticos. Ciencia y cultura, arqueología subacuática [The Mayan underworld and karst systems. Science and culture, underwater archeology]. Instituo Nacional de Antropología e Historia INAH [Mexico's National Institute of Anthropology and History]. Retrieved on November 9, 2021 from: https://sds.yucatan.gob.mx/cenotes-grutas/documentos/el-inframundo-parte1.pdf

Romero-Infante JA (2006) Los mayas, lecciones para el desarrollo latinoamericano [Mayans, lessons for Latin American development]. Cuadernos Latinoamericanos de Administración [Latin American Administration Notebooks], vol II(2):101–112, january-june 2006, University El Bosque (Colombia), ISSN 1900-5016

SARH-IMTA (1988) Las comunidades mayas de Yucatán pasado y presente [The Mayan communities of Yucatan past and present]. 1st ed. Secretaría de Agricultura, Recursos Hidráulicos [Ministry of Agriculture and Hydraulic Resources], Instituto Mexicano de la Tecnología del Agua [Mexican Institute of Water Technology]. Mexico, p 56. Retrieved on November 9, 2021 from: http://repositorio.imta.mx/bitstream/handle/20.500.12013/816/IMTA_003.pdf?sequence=1&isAllowed=y

Schmitter-Sotto JJ, Escobar-Briones E, Alcocer J, Suárez-Morales E, Elías-Gutiérrez M, Marín IE (2002) Los cenotes de la Península de Yucatán [Cenotes in the Yucatan Peninsula]. In: De La Lanza-Espino G, García-Calderón J (eds) Lagos y presas de México [Lakes and dams of Mexico]. AGT Editor, Mexico, pp 337–381. ISBN 978-607-7551-14-0

Terán S, Rasmussen C (2009) La milpa de los mayas. La agricultura de los Mayas prehispánicos y actuales en el noreste de Yucatán [The Mayans' milpa. Agriculture of pre-Hispanic and present-day Mayans in northeastern Yucatan] Mexico's National Autonomous University, Peninsular Center for Humanities and Social Sciences, 2nd edn. UNAM, Mexico, p 395. ISBN 970-607-0206-86-3

Turner BL (1981) Agricultura y desarrollo del estado en las Tierras Bajas Mayas [Agriculture and state development in the Maya Lowlands]. Estudios de la Cultura Maya, vol 13. https://doi.org/10.19130/iifl.ecm.1981.13.546. Retrieved from: https://revistas-filologicas.unam.mx/estudios-cultura-maya/index.php/ecm/article/view/546/541

Wells C (2016) Recent discovery adds to our understanding of ancient Maya civilization. The Houston Museum of natural science. Beyond Bones. Retrieved on November 12, 2021 from: http://blog.hmns.org/2016/09/recent-discovery-adds-to-our-understanding-of-ancient-maya-civilization/

Zapata-Peraza RL (1989) Los chultunes. Sistemas de captación y almacenamiento de agua pluvial [The chultunes. Rainwater collection and storage systems]. 1st edn. Instituto Nacional de Antropología e Historia [National Institute of Anthropology and History], Colección Científica 182, Serie Arqueología [Scientific Collection 182, Archeology Series], Mexico. ISBN 968-60-6874-0

Chapter 4
Water Management in Ancient Mayan Cities

Adriana A. Zuniga-Teran

Abstract Ancestral cultures offer both positive and negative lessons to contemporary societies that are being challenged by growing urban populations and dwindling natural resources. Water is the most critical resource that is essential to human life, therefore understanding the effects of different water management approaches employed by past civilizations is important. Through mostly centralized water management approaches, ancient Mesomerican Maya city-states managed to withstand not only the dry seasons, but also prolonged periods of drought for millennia. Some ancient Maya cities show a high degree of technological achievement in water infrastructure systems, where water was channelized, conveyed, filtered, stored, and distributed to the people or used for agriculture. The Maya cities show sophisticated urban infrastructure systems for water management that we can draw from. However, the Maya also changed land cover and deforested the surrounding landscape, affecting the hydrologic cycle, and consequently, diminished water supplies. This pressure on ecosystems along with prolonged drought and social and political factors caused severe water shortages, starvation, and eventually, migration out of the cities. The Mayan water management experience highlights the need to conserve natural landscapes and ecosystem function to protect the hydrological cycle and the sustainability of our cities.

Keywords Water management · Mayan cities · Water and urban design · Water and power · Natural resources governance

4.1 Introduction

In this chapter, I draw from a literature review to argue that water played a central role in Maya culture because it profoundly influenced their religious, socioeconomic, and political systems, and was a key element of urban design and architecture. To support this statement, I examine water management in ancient Maya cities. I start

A. A. Zuniga-Teran (✉)
University of Arizona, Tucson, USA
e-mail: aazuniga@arizona.edu

G. A. Corzo Perez et al. (eds.), *Aqua Vitae*, The Latin American Studies Book Series,
https://doi.org/10.1007/978-3-031-57731-4_4

by analyzing the water-related challenges experienced in the geographic region, and discuss Maya religious perceptions of water, the critical role that water played in Maya political systems, and how these perceptions influenced urban design and water infrastructure systems. I describe some examples of water infrastructure systems in cities, focusing on Palenque and Tikal as case studies. Finally, I explore some theories on how natural and anthropogenic water scarcity caused the demise of the Maya urban population, and ultimately the collapse of their marvelous cities that marked the end of the classic period.

4.2 Water Challenges for the Maya

The Maya civilization settled in today's Central America and Mexico, mostly in Guatemala and Belize, the western part of Honduras and El Salvador, and the Mexican states of Chiapas, Tabasco, and the Yucatan peninsula (Coe 1993). This civilization flourished between 2000 BCE to 1500 CE, reaching its peak around the sixth century of this current era. The study of this ancient civilization is commonly divided into three periods: (i) Pre-classic (approx. 2000 BCE to 250 CE), (ii) classic, (259 CE to 900 CE), and (iii) post-classic (900 CE to 1542 CE)[1] (Price et al. 2019). The classic period is when the largest cities flourished, including Tikal, Palenque, Copan, Rio Bec, Bonampak, and Calakmul (Fig. 4.1). A combination of natural and anthropogenic factors led the Maya to abandon their cities around the year 900 CE, and this marked the end of the classic period. Other important Maya cities developed in other periods. For example, from the pre-classic period, we can find El Mirador; and from the post-classic period, Chichén-itzá (Lucero 2008). Similar to most indigenous civilizations in the Americas, the demise of the Maya civilization at the end of the post-classic period was mainly influenced by the arrival of the Spaniards. However, the Maya as an agrarian society that shares the same worldview as their ancestors did 4000 years ago still inhabit the highlands-mountainous region in present-day Guatemala and the southern and northern lowlands region in the Yucatan Peninsula (Estrada-Belli 2010).

Since the nineteenth century, scholars have been trying to understand the reasons behind the radical migration away from urban areas at the end of the classic period. One critical factor that explains massive migration is linked to water. Severe water scarcity and subsequent starvation led people to migrate in search of the vital resource. Scholars believe that Maya migrants found refuge in cities in the lowlands of the Yucatan Peninsula. However, water management in the Maya lowlands was also challenging because precipitation in the lowlands is seasonal with extended dry periods that span from December to May (French et al. 2012). In addition, this area is geologically a marine shelf made of limestone that forms a karst plateau (Rice

[1] Scholars no longer consider the time divisions for the Maya civilization (pre-classic, classic, and post-classic) "as stages of cultural evolution or devolution, but simply as arbitrary time divisions that are too engrained in our scholarly literature to be changed" (Estrada-Belli, 2010: 1).

Fig. 4.1 Approximate location of some ancient Maya cities (Coordinates not provided). *Source* A Zuniga-Teran

et al. 2018), which limits the amount of rainfall that can be retained in the soil and infiltrated into the ground resulting in scarcity of surface water during the dry season. In spite of these water availability challenges—along with a lack of beasts of burden that could have helped with construction labor, and the reliance on Stone-Age tools—the Maya city-states in the lowlands managed to flourish for several hundred years (Weaver et al. 2015).

4.3 Religious Perception of Water and Urban Design

The lack of a consistent and reliable water supply influenced Maya religious, political, and social systems, including the design of their cities. The religious conceptualization of water had a strong association with urban design. One of the Maya stories of how the world began involves an ocean and the world emerging from it. The earth was imagined as a turtle or crocodile submerged in water across space and time dimensions. Not only the origin of the world was associated with water, but the origin of the human species was too. The Maya believed that the first human couple came out of a spring in a water cave at the base of a mountain. They regarded their rulers as direct descendants of this first couple, also known as the *First Mother and First Father* (Grauer 2021; Weaver et al. 2015).

Interestingly, Maya cities replicate these religious notions of the world's creation in the layout of their cities—the pyramids represent the mountains, the plazas are equivalent to the ocean or water pools, and the city center symbolizes the water cave. This duality between religious beliefs and architecture is also evident in Maya vocabulary. The same word – *nab* – is used to mean both concepts "pool of water" and "plaza," the same word – *witz* – for both "pyramid" and "mountain," and the same word – *ch'een* – for both "cave" and "city center" (Grauer 2021; Weaver et al. 2015). This way, Maya urban design reflects the culture's cosmic beliefs, and water is a prominent element of both the cities and the cosmology.

Water for the Maya had a symbolic and a practical value. The symbolic value (mentioned above) is described in the *Popol Vuh,* a hieroglyphic book that depicts the world's creation as a site where the springs emerge from the mountains. The practical value of water is, of course, the sustenance of life. *Cenotes*, or sinkholes, and other openings in the earth were considered entrances to the underworld or passageways between the non-human and human worlds (Grauer 2021). Scholars believe that some Maya cities such as Nixtun-Ch'ich' were intentionally located close to a cenote. In Nixtun-Ch'ich', the cenote is at the epicenter of the city with a symbolic centrality not only spatially, but also in time, as this pivotal place is where the community celebrated equinoxes, sunrises, and sunsets, and conducted cosmological rituals. In this city, the cenote is also central to urban planning, as it dictates the main axes of the city (Rice and Pugh 2017). Another example of a city located close to a cenote is Chichén-itzá, which flourished during the post-classic period. The *Cenote Sagrado* (Sacred Cenote) is located 300 m north of the city, and it was a pilgrimage destination for the Maya as well as a site for religious rituals including human sacrifices (Price et al. 2019).

The way the Maya conceptualized the water cycle is also connected to urbanism and architecture. They believed that hollowed mountains were the place where clouds were formed, and these were released to the sky from caves. This imagined view of the hydrological cycle is represented in cities, either naturally or artificially: water in cities flows from springs emerging in the caves of nearby mountains or from human-made pyramids; surface water follows natural streamlines or human-made causeways; and lakes, sinkholes, and other natural water bodies retain water just like their reservoirs, *aguadas,* or *chultuns* did during the dry season (Weaver et al. 2015). In addition, there was a hierarchy in the settlement of urban residents that resembles their conception of the creation of the world. Just as the First Mother and First Father emerged from the water cave, the earlier urban settlers lived close to the water sources, while the newcomers inhabited the peripheries, sometimes without a direct connection to water (Weaver et al. 2015). This settlement pattern happens in most human settlements, regardless of their cosmological beliefs; the distribution of key resources, such as water, dictates how people settle across the landscape. When the resource is concentrated (e.g., reservoir, lake), people gather around it to get frequent access, but when the resource is dispersed, the population settles in a scattered manner. This latter condition presents a challenge for rulers, who have to devise ways to attract people and convince them of their power over the availability of the vital resource (Lucero 2008).

4.4 Water and Political Power

The Maya recognized the pragmatic value of water to sustain life, and this realization positioned water as a symbol of both political and religious power. In most Maya city-states, political power was centralized and sustained by a ruler who provided access to water and thereby the ability to grow food for their people. The Maya rulers-controlled access to water through water infrastructure systems (reservoirs, canals, and aqueducts). Their divine power was legitimized by providing water during the dry season. However, during prolonged drought conditions, this divine ability was difficult to maintain, and in some unfortunate dry times, rulers could lose their political power. When a ruler could not deliver water to a thirsty and starving population, people "voted with their feet," that is, they abandoned the city-state searching for another city-state that could sustain their lives (Grauer 2021; Lentz et al. 2018; Lucero 2008).

Because political power was linked to water access, construction of large-scale water infrastructure in cities was a very important endeavor. Urban water infrastructure was a way for rulers to assert their control of both the metaphysical and biophysical aspects of water. In Tikal and Calakmul, for example, water resources, and economic systems were controlled by the cities' divine rulers, who were responsible for providing and maintaining water infrastructure systems, and performing sacred water rituals based on symbolism and cosmology (Grauer 2021; Lucero 2008). Feasts and ceremonies were offered by Maya leaders at plazas, temples, and ball-courts, to unify people emotionally, spiritually, and physically, while establishing the prestige of elites (Lucero 2008). Not only the construction of water public works required advanced engineering knowledge, but their operations and maintenance required sophisticated know-how and continuous use of resources too. Disposing of human waste and ensuring drinking standards for standing water in reservoirs during the dry season were capital-intensive endeavors (Lucero 2008).

Although most of the Maya city-states had centralized governments, most scholars believe that there may have been other, more egalitarian political systems, and this assumption is based on archeological evidence of decentralized water infrastructure systems. A communal decentralized management system would give water access to all residents of the city, without the control of a ruler (Marken et al. 2019). For example, research has found that rainwater harvesting may have been practiced at the household level in the Maya city of Caracol in Belize, undermining the assumption of political control through water management in all Maya city-states (Chase 2016). Another example is the city of Aventura, a medium-size Maya city in northern present-day Belize, where access to water cut across all social classes. Both wealthy and low-income households in Aventura show pocket *bajos* or "non-draining areas of low elevation, which are up to 2 km^2 in area and up to 3 m deep" that might have functioned as water storage infrastructure at the household level (Grauer 2021: 2). Access to bajos in Aventura was possible due to the presence of steps. This characteristic attracted people from other cities during periods of drought in the classic period (Grauer 2021).

4.5 Maya Urbanism and Water Management

It is very difficult to understand the relationship between urbanism and water management because the current conditions of the sites do not permit them to function as intended. Nevertheless, learning the hydrologic system of Maya cities has been possible through the use of remote sensing, spatial analysis, and hydrologic modeling (Weaver et al. 2015). Scholars tend to agree that Maya urbanism served political and religious purposes, but it also had a critical utilitarian purpose: to control the flow of water and retain the precious resource during the dry season (Weaver et al. 2015).

Box 4.1 Water Infrastructure in Palenque

Palenque, a small Maya city located in the mountains of Chiapas, Mexico, has one of the most fascinating water infrastructure systems of ancient civilizations (Fig. 4.2). The city was located close to numerous water sources. According to Stuardo (2020), there were at least 56 natural water sources near Palenque, and this is probably the reason one of the glyphs for Palenque was "Lakamha," which means "Great Waters." Its location also made sense with respect to fertile adjacent lands for agriculture (Stuardo 2020). In addition, the location of this city was strategic for disputes with other city-states. Palenque was built in a steep and small space defined by topography, which caused high-density development and allowed the community to spot invaders from far away. The hilly terrain also discouraged invaders to pass through this city. The spatial constraints of Palenque—a confined plateau—allowed residents to capture water from nearby springs, but the terrain also complicated the construction of buildings and their maintenance during the rainy season (French et al. 2012).

Fig. 4.2 Palenque, Chiapas, Mexico. *Source* Bernard Dupont (public domain)

Palenque faced flooding and erosion problems, as well as the division of public space by the creeks. To address these challenges, builders channeled the creeks in small aqueducts that ran underneath the plazas (French et al. 2012). The people of Palenque likely expended massive labor in the building of their water infrastructure system and its maintenance, which sometimes required the flattening of slope surfaces into artificial terraces to direct runoff to desired streams of canals. This amazing engineering work was necessarily coordinated and directed by experienced individuals who exerted social control (Stuardo 2020).

Palenque's water infrastructure system consists of hydraulic works that direct, channel, or contain water. These works include aqueducts having diverse capacity levels and varying complexity. For example, archeologists have found drains that control water excess flow during the wet season, and others that conserve water during the dry season. There are also bridges that permitted people to transit over the aqueducts, pools, canals, and even small rock dams that changed water flow (Stuardo 2020).

Water pressure was also mastered by the Maya in Palenque. The canals were narrower as they approached the river, which makes scholars believe that the Maya used some type of gating system to control the flow in the canals and thereby practice year-round agriculture. Local agricultural practices adapted to the hilly terrain too. Channelized farming fields were built to deal with both flooding and drought in Palenque. This infrastructure served two functions: it drained the field during the wet season, and irrigated the land through drainage irrigation during the dry season. This type of irrigation controls the water table levels on the canals and in the field (French et al. 2012).

An issue with the channelized agriculture system in Palenque and a megadrought might have influenced the eventual abandonment of the city in 799 CE. However, scientists believe that perhaps drought refugees seeking refuge in Palenque from other places might have strained the hydrological system. They also think that land cover change in Palenque might have affected the hydrological cycle and water flow at the watershed scale and, consequently, played a role in the demise of the population (French et al. 2012).

The emergence of city-states in the Maya lowlands during the classic period was made possible by water storage infrastructure. The lowlands area (southern Mexico, Guatemala, and Belize) lacks access to surface water bodies (rivers or springs). Therefore, to sustain a population during the several-months-long dry season, the Maya had to build reservoirs fed by drainage ditches (Bower 1991). Rainwater was harvested and conveyed to large central stone-lined reservoirs, which were coated with clay or plaster. In addition, there is evidence of small reservoirs at the household level, as described above. Scientists have found remains of some 1600 household reservoirs in the City of Caracol in Belize. Using remote sensing data to map these reservoirs, they located more than 25 times the number of reservoirs identified by archeological findings (Chase 2016).

Urban water infrastructure systems allowed Maya populations to withstand periods of drought, and the Maya used several types of water storage infrastructure (Table 4.1). One kind of rainwater harvesting infrastructure found in Maya cities was the *aguada* landscape feature, which was either natural or human-made and consisted of a pond that stored rainwater. Aguadas apparently played an important role in the fate of lowlands cities. Natural aguadas result from collapses on the terrain or sinkholes and were typically lined with stones and coated with clay to function as reservoirs. Human-made aguadas were usually abandoned quarries transformed to store water (Ferrand et al. 2012).

Most Maya cities in the lowlands are located near a *bajo*, a critical landscape feature for survival during the dry season. Calakmul and Tikal, for example, are cities located in well-drained and low-relief areas. These cities grew following the terrain and the possibility to capture rainwater in "still water systems," i.e., bajos (French et al. 2012). Archaeologists debate whether bajos might have been wetlands or lakes during the classic period or might have served as agricultural fields for

Table 4.1 Different types of water storage infrastructure found in Maya cities

Water storage infrastructure	Description	Reference
Aguadas	Small natural or human-made ponds	(Ferrand et al. 2012)
Bajos	Broad flat depressions, or low-lying wet terrain	(Dunning et al. 2015)
Reservoirs	Abandoned quarries transformed into large impermeable water storage basins	(Weaver et al. 2015)
Communal/district tanks	Larger-capacity storage structures linked to a hydrological infrastructure system	(Marken et al. 2019)
Rejolladas	Deep karst sinkholes	(Lentz et al. 2018)
Chultuns	Underground storage chambers	(Weaver et al. 2015)

the Maya. The specific use of the bajos, however, is context specific and cannot be generalized (Dunning et al. 2015).

Similar to the origin of some aguadas, the basis of reservoirs may have been the quarries that the Maya used for their main building material in adjacent structures. Reservoirs were also coated with plaster that made them more impervious than the surrounding karst landscape. In some cases, the slope of the plazas was intentionally designed to direct runoff to the reservoirs (Weaver et al. 2015). Reservoirs often experienced erosion in the embankment because they were very steep. So, the Maya periodically drained and dredged the reservoirs to maintain their capacity (Lentz et al. 2021).

The banks of the reservoirs were bordered with tropical vegetation used to stabilize the embankments. This way, water and vegetation were used complementarily to provide ecosystem services and a year-round water supply. Tikal has one of the largest remaining systems of earth and stone reservoirs in the Pre-Columbian Americas (Lentz et al. 2021) (See Box 4.2). Some reservoirs have steps or ramps for easy access during different seasons when the water level fluctuated. The easy access from the surrounding areas suggests that everyone in that city had access to water in that specific reservoir (Marken et al. 2019). However, in some cities, a system of gates could have controlled water flow upstream. Interestingly, analogous types of rainwater harvesting infrastructure systems were also developed by ancient civilizations in India, in the form of stepped ponds (*jhalaras*), or stepped wells (*baoris*), and around similar eras (eighth century, CE) (Vyas 2011).

There is evidence that the Maya developed a purification filtration system for water that used sand to clean the water in the reservoirs (Tankersley et al. 2020; Scarborough et al. 2012). These filtration systems may be the earliest instances anywhere of aquaponic practices. Findings have shown that water stored in the reservoirs of Maya cities must have included fish, turtles, and aquatic plants to ensure the quality of the standing water. In addition, field researchers have found a layer of pots located at the bottom of reservoirs. The pots might have held the roots of aquatic plants, such as waterlilies (Lentz et al. 2021).

Some extant Maya cities display an interconnected, three-stage-hierarchy water infrastructure system. For example, El Perú-Waka' is a Maya city located in the Petén area that features four public plazas, monumental structures, medium-size reservoirs, and a system of canals that distributed water to the peripheries. El Perú-Waka' has three main types of reservoirs: (1) domestic aguadas, (2) high-status residential reservoirs, which are medium-size works found near elite residential areas, and (3) communal or district tanks, which have the largest capacity and are found in communal areas of the city. There is an interconnection between the three types of reservoirs in this city. Urban water management in El Perú-Waka' shows similarities with cities in Arequipa, Peru, which also has a three-stage water management hierarchy for irrigation, and where urban infrastructure is used to direct runoff to the reservoirs (Marken et al. 2019).

Box 4.2 Water Infrastructure in Tikal

Tikal was one of the most important Maya cities that lasted 1500 years, located at the epicenter of the Yucatan peninsula, in modern-day Guatemala. Tikal grew during the late classic period (600–850 CE). From the third century to the middle of the ninth century, its population quadrupled, peaking in 830 CE with 6,2000 residents (Lentz et al. 2021). It was relatively densely populated even though there was no source of water nearby, a characteristic made possible only because of the comprehensive planning of the functional water infrastructure system (Fig. 4.3) (Lentz et al. 2018). Tikal is an example of sophisticated hydrologic engineering implemented by a hierarchical political system (Lentz et al. 2021). The springs, canals, and different reservoirs are all integrated into a holistic system, which includes the largest known dam in the region, a cofferdam for dredging it, a switching station that allowed the release of waters from the reservoirs, a series of elevated rainwater catchment areas that directed runoff to the reservoirs, and a deep canal system made of rock to convey water (Fig. 4.3). The designers of Tikal also built elevated silting tanks, which expanded the head end of the creek and discharged water into the deeper reservoir (Scarborough et al. 2012).

Fig. 4.3 Reconstruction model of Tikal. *Source* Museo Nacional de Arqueología y Ethnología (National Museum of Archeology and Ethnology), Guatemala (public domain)

The Maya paved the hilltops that surrounded the canyons to direct runoff to the reservoirs. Within the city, they used the slope of the plazas and courtyards to direct runoff to reservoirs and even the ceremonial precincts were integrated into the water infrastructure system (Lentz et al. 2018). The reservoir system consisted of six paved drainage systems feeding about 90,0000 m^3 of water every year into 10 reservoirs located in the center of the city. Water was then released through a controlled system into smaller storage basins located at the peripheries of the city. These basins might have been used for irrigation, sustaining year-round agriculture. Central Tikal also reveals the existence of residential water storage structures used by individual households, but these do not seem to have been connected to the central reservoir (Bower 1991). Some scholars believe that the population of Tikal may have reached 80,000 residents around 750 CE, mainly because their storage water system attracted people to the city (Bower 1991).

Tikal also shows a sophisticated water purification system that used a mixture of zeolite and quartz. Zeolite removes pollutants and is non-toxic, porous, and naturally adsorbent. It also disperses insoluble and soluble toxins, making water potable enough for drinking (Lentz et al. 2021; Tankersley 2020).

It is unlikely that reservoirs experienced sedimentation problems due to the paving of plazas and the presence of a bench that acted as barrier for sediment. The Maya used vegetation to hold this bench and prevent soil erosion. They probably used perennial plants with deep roots, in combination with fruit trees (e.g., cacao, avocado, native trees). In this way, reservoirs served as artificial lakes surrounded by orchards, providing aesthetic value that was appreciated by the elites, who mostly resided near palaces. Water lilies may have been used to decrease evaporation of water in the reservoirs during the hot days. Water lilies were also symbolic for the Maya; they are often depicted in their murals (Lentz et al. 2021).

Tikal did not have access to permanent water sources, except for a spring that provided water during the pre-classic period. Unreliable water sources rendered the residents of Tikal vulnerable to drought, which they were able to withstand thanks to their advanced hydraulic system. However, extensive paving and land cover change likely modified the water cycle, limited the recharge capacity of the land, and ultimately decreased the flow of the springs. This effect on the water cycle increased the population's dependence on rainfall (Lentz et al. 2014, 2018). In addition, evidence suggests that Tikal's reservoirs became contaminated with mercury and developed periodic toxic algae bloom (Lentz et al. 2021). A vicious cycle of land cover change, prolonged drought, pollution, and desertification likely caused starvation, loss of faith in the ruler's divine power, and subsequent mass migration away from Tikal in search of other cities. Because the last monument built in Tikal dates to 869 CE (Lentz et al. 2014), the scholarly consensus is that people abandoned Tikal in the ninth century (Lentz et al. 2020).

4.6 Beyond Cities: Maya Agriculture and Water Management

Agricultural practices are a fundamental aspect of water management today, as they were in ancient societies, because agriculture is the largest water user; globally, around 80% of the water in a basin is used for agriculture (Scott et al. 2015). Accordingly, when analyzing historic water management approaches, one can safely assume that agricultural practices played a major role in water usage. Maya cities in the lowlands functioned as low-density agrarian communities, similar to cities in southeast Asia, such as Angkor Wat, that have similar hot and humid climates with marked wet and dry seasons (Marken et al. 2019). Maya cities were supported by a large population of farmers, who practiced three main types of agriculture. First, they might have grown crops using primitive "slash-and-burn" techniques, which involves the cutting and burning of plants in forests and is a hallmark of tropical rainforest regions. This practice, however, quickly decreases the soil fertility, forcing farmers to move elsewhere within a few years of farming—4 to 7 years in highlands like Peten, and 10 to 15 years in lowlands like Yucatan. Second, the Maya might have used the raised field system, which consisted of narrow rectangular plots raised above the bajos, harboring fish, and reptiles. Bajos were found mostly in Campeche and Quintana Roo states in Mexico, and in Belize, covering 40% of the landscape. Finally, the Maya also grew crops on stone-walled terraces, which functioned as silt traps in hilly terrains. This practice allowed them to grow sufficient quantities of food to sustain a large population (Coe 1993; Sharer 1996).

Some cities deployed a diversity of agricultural practices. Tikal citizens, for example, practiced a variety of agriculture systems that ranged from household gardens to root-crop agriculture, arboriculture, irrigated fields, short-fallow cropping systems, and bajo margin cultivation (Lentz et al. 2014, 2018). They used *rejolladas* (deep karst sinkhole), terraces, reservoirs, and canals to retain rainwater for both drinking and irrigation (Lentz et al. 2018). The Maya farmed the periurban bajos areas of Tikal with maize and root crops (Lentz et al. 2021). Terminos was a suburban settlement in the outskirts of Tikal that used terraces made by chert debris to retain soil moisture (Lentz et al. 2014).[2] The growers in Terminos managed their agriculture locally. This may be the reason why Terminos outlasted Tikal until the Post-classic period and their aguada continues to function today (Lentz et al. 2018).

4.7 Deforestation and Migration

Evidence suggests that the Maya systematically deforested the land surrounding their cities. One of the reasons was their need for stucco in monumental stone building. Stucco is quicklime that when combined with water results in a fine plaster for the

[2] Chert is a stone used by the Maya for tool manufacture. A large chert-bearing zone lies in northern Belize (Shafer and Hester, 1983).

construction of buildings, plaza floors, and aqueducts. To make stucco, limestone must be burned at very high temperatures (900 °C), requiring substantial quantities of wood (French et al. 2012), which drove deforestation.

Slash-and-burn agriculture was another practice that led to deforestation around some Maya cities. In Tikal, for example, people engaged in this type of agriculture during the pre-classic period. As noted, when vegetation is removed, soils lose their nutrients and eventually erode. Apparently, the Maya learned from this mistake, and during the classic period, they managed their forests more sustainably. They left some areas of the forest rest (fallow) so that the soil could recover its nutrients (Lentz et al. 2018). Nevertheless, unsustainable agricultural practices and uncontrolled logging and loss of vegetation likely affected the hydrological cycle and influenced the subsequent decrease in rainfall and decrease in agricultural productivity that drove the population's collapse (Lentz et al. 2018; Scarborough et al. 2012).

The classic period witnessed a series of downturns and recoveries in Maya urban population. Some cities recovered quickly from abandonment, while others did not. Research suggests that cities in elevated areas were more likely to collapse than those in lower-lying areas (Dunning et al. 2012). Water storage capacity in the lowlands region attracted migrants, just as an oasis attracts wildlife in a savanna, resulting in population growth (Bower 1991). The karst physiography of the lowlands allowed the retention of seasonal water and this land characteristic allowed the flourishing of the Maya for several additional centuries (Scarborough et al. 2012). However, mass in-migration increased water demand, and this could have strained the water infrastructure system, as it did in Palenque (French et al. 2012).

In addition to migration and deforestation, the inability to store food (the Maya did not have granaries) likely played a role in the demise of the population of Tikal and other Maya cities. Food storage was small in scale, likely done at the household level, with a surplus of no more than three years. Therefore, to some extent, the Maya were able to endure water scarce periods by supplementing agriculture with trade. They used cacao beans as currency to be used to buy food during periods of drought. But after a few years or severely diminished agricultural productivity, stored food, and accumulated currency would run out, and desperate residents would seek sustenance elsewhere. In addition, a longer-than-usual prolonged drought likely triggered a destruction of social order, lack of faith in the political leader's divine legitimacy, and an imminent collective exodus (Grauer 2021; Lentz et al. 2018). As stated by Lucero et al. (2011: 479), "Maya farmers are still around today; kings, however, disappeared over 1000 years ago."

4.8 Conclusion

The ancient Maya civilization was deeply reliant on and associated with complex and sophisticated water infrastructure systems and centralized water management. Even though the tropical region of Mesoamerica is not considered water scarce, the stress caused by annual dry seasons likely marked a spiritual conceptualization of the universe. This outlook and world view strongly influenced Maya political and socioeconomic systems. Rulers who were able to provide water during the dry season were legitimatized as divine. This led to an intrinsic incentive to construct complex water infrastructure systems in Maya cities—a task that was only possible under centralized government systems. Although there exists some archeological evidence of decentralized water management approaches in a few Maya cities, centralized approaches likely dominated water management for the Maya.

Urban design reflects Maya centralized water management approaches and displays water-related cosmic and religious beliefs. Some still-standing, complex urban water infrastructure systems can still be admired today. Tikal and Palenque are fine examples of sophisticated engineering solutions to water management challenges that enabled cities to sustain large populations for over a thousand years.

However, the Maya civilization experienced a combination of water-related factors that influenced their demise. These factors include megadrought conditions, increased water demand (due to immigration), reservoir water pollution, desertification of surrounding landscapes, and land cover change. All these factors affected the hydrological cycle, reduced agricultural productivity, and eventually caused starvation, which in turn eroded faith in the ruler's ability to provide water. It appears that desperate surviving populations abandoned their cities to seek water, safety, and spiritual refuge elsewhere. A key sustainability and water management lesson from the Maya is the need to protect the hydrological cycle by conserving natural landscapes and ecosystem function within and beyond cities.

References

Bower B (1991) Water storage spurred growth of Maya cities. Sci News/ Soc Sci Public 139(6)

Chase ASZ (2016) Beyond elite control: residential reservoirs at Caracol. Belize 3:13

Coe M (1993) The maya (Fifth Edition). Thames and Hudson

Dunning NP, Beach TP, Luzzadder-Beach S (2012) Kax and kol: collapse and resilience in lowland Maya civilization. PNAS 109(10):3652–3657

Dunning NP, Griffin R, Jones J, Terry R, Larsen Z, Carr C (2015) Life on the edge: Tikal in a Bajo Landscape. In: Lentz DL, Dunning NP, Scarborough VL (eds) Tikal: paleoecology of an Ancient Maya City. Cambridge University Press, pp 95–123

Estrada-Belli F (2010) The first maya civilization: ritual and power before the classic period (EBook). Taylor and Francis Group

Ferrand EA, Dunning NP, Lentz DL, Jones JG (2012) Use of aguadas as water management sources in two southern Maya lowland sites. Ancient Mesoamerica Cambrid Univ Press 23(2):85–101

French KD, Duffy CJ, Bhatt G (2012) The hydroarchaeological method: a case study at the Maya site of Palenque. Lat Am Antiq 23(1):29–50

Grauer KC (2021) Heterarchical political ecology: commoner and elite (meta)physical access to water at the ancient Maya city of Aventura, Belize. J Anthropol Archaeol 12

Lentz DL, Dunning NP, Scarborough VL, Magee KS, Thompson KM, Weaver E, Carr C, Terry RE, Islebe G, Tankersley KB, Grazioso Sierra L, Jones JG, Buttles P, Valdez F, Ramos Hernandez CE (2014) Forests, fields, and the edge of sustainability at the ancient Maya city of Tikal. Proc Natl Acad Sci 111(52):18513–18518. https://doi.org/10.1073/pnas.1408631111

Lentz DL, Dunning NP, Scarborough VL, Grazioso L (2018) Imperial resource management at the ancient Maya city of Tikal_ A resilience model of sustainability and collapse. J Anthropol Archaeol 52:113–122
Lentz DL, Hamilton TL, Dunning P, Scarborough VL, Luxton P, Vonderheide A, Tepe EJ, Perfetta CJ, Brunemann J, Grazioso L, Valdez F, Tankersley KB, Weiss AA (2020) Molecular genetic and geochemical assays reveal severe contamination of drinking water reservoirs at the ancient Maya city of Tikal. Sci Rep 10(10316):9
Lentz DL, Hamilton TL, Dunning NP, Tepe EJ, Scarborough VL, Meyers S, Grazioso L, Weiss A A. (2021) Environmental DNA reveals arboreal cityscapes at the Ancient Maya Center of Tikal. Sci Rep 10
Lucero LJ (2008) Water control and Maya politics in the Southern Maya lowlands. Archeol Pap Am Anthropol Assoc 9(1):35–49. https://doi.org/10.1525/ap3a.1999.9.1.35
Lucero LJ, Gunn JD, Scarborough VL (2011) Climate change and classic maya water management. Water 3(2):479–494. https://doi.org/10.3390/w3020479
Marken DB, Ricker M, Rivas A, Maxson E (2019) El urbanismo de baja densidad en las Tierras Bajas Mayas: El caso de El Perú-Waka', Petén, Guatemala. PNAS 32
Price TD, Tiesler V, Freiwald C (2019) Place of origin of the sacrificial victims in the sacred Cenote, Chichén Itzá, Mexico. Phys Anthropol 170:98–115
Rice PM, Cordell A, Kidder G, Harris W, Pugh TW, Chan Nieto E (2018) Early construction at Nixtun-Ch'ich', Petén, Guatemala_ An architectural-footing and -bonding sample. J Archaeol Sci 17:754–761
Rice PM, Pugh TW (2017) Water, centering, and the beginning of time at Middle Preclassic Nixtun-Ch'ich', Petén, Guatemala. J Anthropol Archaeol 16
Scarborough VL, Dunning NP, Tankersley KB, Carr C, Weaver E, Grazioso L, Lane B, Jones JG, Buttles P, Valdez F, Lentz DL (2012) Water and sustainable land use at the ancient tropical city of Tikal. Guatemala PNAS 109(31):12408–12413
Scott CA, Kurian M, Westcoat JL Jr (2015) The water-energy-food nexus: adaptive capacity to complex global challenges. In: Kurian M, Ardakanian R (eds) Governing the nexus: water, soil and waste resources considering global change. Springer, pp 15–38
Shafer HJ, Hester TR (1983) Ancient Maya Chert workshops in Northern Belize Central America. Am. Antiq. 48(3):519–543. https://doi.org/10.2307/280559
Sharer R (1996) Daily life in Maya civilization (First). Greenwood Publishing Group, Inc
Stuardo RL (2020) LAKAMHA: THE PLACE OF "BIG WATERS" The archaeology of the ancient city of Palenque, Mexico. S. R., pp 16
Tankersley KB (2020) Zeolite water purification at Tikal, an ancient Maya city in Guatemala. Sci Rep 7
Vyas SP (2011) Regional water management: a case study of some step wells of Jalore region. Proceed Indian Hist Congr 72:1208–1216
Weaver E, Carr C, Dunning NP, Florea L, Scarborough VL (2015) Examining landscape modifications for water management at tikal using three- dimensional modeling with ArcGIS. In: Lentz DL, Dunning NP, Scarborough VL (eds) Tikal: paleoecology of an ancient Maya City. Cambridge University Press, pp 87–94

Part II
Visions and Viewpoints

Chapter 5
Engineering and Veneration of the Yacumama (Mother Water) in Inca Culture

Alberto Huiman Cruz

Abstract The Inca culture had a cosmology that included the stars (the sun, the moon, and the stars) and natural resources such as water and soil. These elements were considered divine and were respected by the Inca people, who believed that this respect would be reciprocated with abundant food for their community. From here, perhaps, comes the explanation for the impressive hydraulic works that they constructed, which continue to inspire admiration and ongoing research. These works are not only remarkable for their structural achievements, but also for the reverence they inspire. The veneration brought with it spaces of protection because they represented a source of life and food supply. There are seven hydraulic systems that have gained renewed importance in the twenty-first century due to their technological efficiency in water management. These systems are: (1) High Andean dam systems, (2) platform systems, (3) mound systems, (4) irrigation systems, (5) production systems, (6) aqueduct systems, and (7) storage systems. But a vision of preservation is necessary, which should represent a contribution to the fight against poverty under a multidimensional approach: engineering, agro-ecology, environment, and history. To begin this process, it is essential to systematize and consolidate the existing information on the legacy, based on quantitative and qualitative data, highlighting the technical nature of the work and its importance.

Keywords Inca · Engineering · Mother water · Preservation · Hydraulic systems · Veneration

5.1 Introduction

The Inca culture was settled in the Cuzco valley, with urban and rural areas intertwined by numerous irrigation ditches (Sherbondy 1987). Multiple factors, such as altitude, topography, location, and hydrography of the valley within the Andean

A. H. Cruz (✉)
PWI (Peru Waste Innovation), San Isidro, Perú
e-mail: alberto@pwi.com.pe

G. A. Corzo Perez et al. (eds.), *Aqua Vitae*, The Latin American Studies Book Series,
https://doi.org/10.1007/978-3-031-57731-4_5

mountain ranges, were key to the initiative of humans to establish their homes and make decisions about modifying them (Beltrán 2013). All this was favorable for the establishment of the Inca culture and the full development of their activities, both agricultural and livestock, because they had flat spaces, fertile soils, and water resources (Miño 1994).

It is the latter, the water resource, which had great presence in the geography of the Cuzco valley, either in the form of springs, rivers, irrigation canals or the abundant rainfall characteristic of the place, being of vital importance in the Inca culture (Miño 1994). Based on this, the Inca culture developed a solid cosmovision of water, which was part of their divinities and to which they owed respect. This is probably the explanation for the great hydraulic works that they left and are reasons for admiration and constant research, not only for the structure but also for the veneration (Ancajima 2016).

The hydraulic systems developed by the Inca culture are an example of technological efficiency for water management, and in the twenty-first century, they are revalued; however, a vision of preservation is necessary, which should represent a contribution to the fight against poverty under a multidimensional approach: engineering, agro-ecology, environment and history. To start this process, it is essential to systematize and consolidate the existing information on the legacy of our ancestors, which obeys quantitative and qualitative data, highlighting the technical nature of the work and its importance for development (Ancajima 2016).

5.2 The Cosmovision of Water

The Inca empire maintained an Andean worldview that emphasized people's appreciation of their cultural and natural environment, resulting in an evolutionary process of thought (Cruz 2018).

Since ancient times, our Inca ancestors have viewed water as a symbol of fertility, often represented by spirals and vortexes. In some cultures, the Nazca lines have been venerated as they reflect stories related to deities associated with water. For example, consider the case of the god Kon, whose myth involves populating the earth with humans and providing them with abundant water and fruits. Similarly, the observation of the Milky Way, named Mayu, was deemed important in predicting the water cycle during the solstices (Ribeiro 2013).

Our ancestors also worshiped through ceremonies. Such is the case with the well-known ushnus, where water was worshiped as an element, particularly the gods who had control over its various forms such as rain, rivers, seas, and lagoons. Through the ushnus, the gods performed three functions: bringing water from the world above (Hanan Pacha), materializing or acquiring a visible form in the middle world (Cay Pacha), and fertilizing the Pachamama by entering the world below (Monteverde 2016, p. 249). Within the valley, there were numerous irrigation ditches integrated into various hydraulic systems or networks. The sources of these systems were considered sacred and were consecrated in the form of huacas (shrines) to which offerings were made (Sherbondy 1987).

5.3 Water as Social Issues

The divisions of the Inca empire's territory and administration were four: Collasuyo, Antisuyo, Chinchaysuyo, and Contisuyo. This division does not correspond to four homogeneous territories. This is because, within each pair of opposite but complementary elements, such as left/right or up/down, one is superior to the other. In their respective situations, Chinchaysuyo and Collasuyo were superior to Antisuyo and Contisuyo, respectively. Similarly, Chinchaysuyo was superior to Collasuyo. Therefore, although the Tahuantinsuyo was not the first process of centralization and political unification in the region, the political organization and administration of the Inca empire were crucial in achieving its maximum state development. Along with its warlike capacity, it allowed the government to bring together the various communities that inhabited the central Andes region (Revenga 2017).

From a political perspective, the Inca empire utilized hydrology in three ways. One of the key factors in land economy is the location of population centers. Another important factor is the use of surface water through hydraulic works. Examples of hydraulic works include headwater projects in rivers or lakes, irrigation canals, aqueducts, water reservoirs, lining of artificial channels, and river channeling walls. These measures not only provide protection to the banks of the river, but also ensure a reliable water supply for agricultural and urban use. Finally, there is the use of groundwater in relation to excavations such as pits and wells, as well as the utilization of springs and construction of drainage systems (Regal 1945).

The Inca empire established clear boundaries for its dual division: Hanan Cusco (upper region) and Hurin Cusco (lower region). The empire also divided the four suyos in Cusco based on the natural hydrology of the Huatanay River, which was the dominant river in the valley. Additionally, the Inca empire established regulations for resource use, including land and water rights (Sherbondy 1987; Miño 1994).

The geography of the natural hydrology and the contours of the areas irrigated by canals became important factors in defining the boundaries of administrative subdivisions during the Inca period. The distribution of water and land rights was coordinated among specific groups whose territories were determined by the principles of the radial organization of the Cuzco valley. For instance, the panaca, a corporate group that identified with the Inca nobility, controlled the most significant irrigation ditches and lands in the valley (Sherbondy 1987).

Finally, in terms of administration, the Inca empire utilized subordination and subjugation to unify their lands. They assimilated the hydraulic works of other organizations that they subdued in the process (Regal 1945). However, an Inca was assigned to build the hydraulic works, and they, in turn, became the founders of their respective communities. In the case of Hurin Cuzco, Sinchi Roca oversaw the construction of drainage works for the swamp or lake located in the central area of

Cuzco. This allowed for the construction of the Aucaypata square, which is now known as the Plaza de Armas. Similarly, Sinchi Roca directed the waters of the Huatanay River, enabling the cultivation of previously unusable land and improving flood control. A similar case occurred in Hanan Cuzco, where Inca Roca constructed irrigation systems to transport water upstream toward the ponds and cultivated fields (Sherbondy 1987).

5.4 Engineering Works

There are seven hydraulic systems that exemplify technological efficiency in water management, and they are being reevaluated in the twenty-first century.

5.4.1 High Andean Dam Systems

In the Andes mountain range, there are several dams that are known as high Andean dams due to their location. These dams have a medium capacity and were built to store rainwater. Capacities ranging from hundreds of thousands to a million cubic meters were highly valued in the past. These dams were designed in a way that did not obstruct the riverbeds, which is in contrast to the current trend of installing large dams directly in the river axis (such as Gallito Ciego). Unfortunately, due to the natural process of basin erosion, these dams are becoming clogged and are only operating at around 50% of their capacity (Ancajima 2014).

Forty reservoirs, constructed with clay and stones, have been discovered in the upper region of the Nepeña River basin in the Cordillera Negra mountain range of the Ancash department. The pre-Incas constructed these reservoirs to regulate water flow in streams, facilitate infiltration in lower areas, and supply springs and wetlands with water. Currently, some of these systems are also utilized for water harvesting, which is subsequently used to irrigate pastures, potatoes, barley, or natural grasslands. However, what is admirable about the technology implemented, in addition to the management of the territory to capture and store water in the upper areas and harvest it in the lower areas, is the longevity of the dams and dikes (Varilla 2015).

In this regard, Llosa (2008) mentions that to date, only 13 out of the 40 dams have received maintenance with the support of some cooperation projects. It is worth mentioning that, generally, dams were constructed in a basin or collecting basin that had a natural depression and an outlet that was dammed with clay, stone, and mud. This process created artificial lagoons. Therefore, some ancient dams were enhanced by incorporating a sand trap or sediment trap, regulating outlet valves, and implementing purges. However, other dams have undergone reconstruction with reinforced or cyclopean concrete.

According to Varilla (2015), the areas where the recovery experiences of this technology are located include the districts of Cáceres del Perú, Moro, Nepeña,

Samanco, and Pamparomás, which are situated in the provinces of Huaylas and Santa in the Department of Ancash. It is presumed that the ancient settlers were able to find suitable locations for water storage and infiltration after several years of trial and error tests, utilizing the hollows in the land.

Therefore, he suggests that rehabilitating the dams for present reuse would yield the following benefits:

(I) There has been an increase in water flow in the springs located in the lower areas.
(II) The restored dams provide increased access to water resources for human and agricultural use.
(III) The recovered dams provide an opportunity to value the water resources and the territorial system, which in turn strengthens the culture and capacity for managing micro-basins. This enables the formation of a united front to address potential impacts generated by climate change.

Therefore, in order to ensure the sustainability of this technology, it is essential to consolidate recovery and improvement processes. This includes maintenance, cleaning, and de-silting of the pond or dam basins, as well as rehabilitation of the fine material that comprises the dam structure.

5.4.2 *Platform System*

The citadel of Machu Picchu is a significant example of water management in the Andean region. However, before the late twentieth-century research conducted by Wright and his colleagues, it was impossible to determine the source of the water that supplied it. However, it is known that in order to meet the population's demand for drinking water, particularly during the dry season (from April to August), the Incas must have first identified the springs that provided water to the city. The drainage system built on the agricultural terraces or platforms and in the central plaza of Machu Picchu is designed with three layers. The top layer contains topsoil, the middle layer is composed of gravel (fine stones and sand), and the deepest layer consists of large stones. Consequently, this design allows some of the rainwater to remain in the top layer of the crop, moistening the roots. Any excess water either evaporates through evapotranspiration or drains slowly into the intermediate layer of gravel before passing through the larger stones toward the interior of the mountain or to sources on lower floors (Ponce-Vega 2015). The hydraulic works found in Machu Picchu reflect that it was one of the Inca's top priorities during its construction. According to Wright and Valencia (2009), 60% of the constructions in the citadel were focused on stabilization works, including the drainage system. "The Incas began constructing temples and dwellings on top of these stabilized sites only later," (Carlotto et al. 2009, p. 736). In a similar fashion, there exist the muyus—four naturally formed holes created by rain and subway currents that have molded the karst stone and shaped four dolines that resemble geometrically regular bowls. These dolines are located in the town of Moray, within the Maras district of the Cusco department (Ponce-Vega 2015) (Fig. 5.1).

To protect the circular terraces or platforms, the Muyus constructed 27 stone channels aligned vertically in one direction. These channels followed an imaginary line that terminated at the sump. For ceremonial purposes, water was transported in an orderly manner from the reservoirs to the stone channels. These channels were directed from terrace to terrace and then infiltrated into the sinkhole (Wright et al. 2011). It is noteworthy that despite four centuries of exposure to a harsh environment, the slopes remain healthy and the bottom of the muyu remains free from salts and erosion.

Kenneth Wright (2011) stated that in Moray, where the muyus are located, modifications were made to the topography through the use of catchment canals, regulation and storage reservoirs, cone-shaped filters, and terraces. These modifications allowed for the stabilization and beautification of the slopes and the bottom of the muyus, making it a masterpiece of environmental art.

The composition and structure of the Andes, which consist of metamorphic, sedimentary, and intrusive rocks, allow water to infiltrate through fractures and diaclases at high altitudes. This water accumulates in small spaces between the rocks and

Fig. 5.1 Muyus, locality of Moray, district of Maras, department of Cusco. *Source* Puquios, qanats and springs: water management in ancient Peru (Ponce-Vega 2015, p. 289,)

emerges at lower altitudes. Wright and Valencia (2009) also mentioned that they discovered the main spring that supplies Machu Picchu is located at an altitude of 2458 m.a.s.l. On a hill to the north of the citadel, it receives rainwater from a tributary watershed that covers an area of approximately 16.3 hectares. Wright et al. (1997) noted that the primary spring generated a flow ranging from 25 to 125 L per minute, depending on the season. During the dry season (April-August), the flow was lower, while during the rainy season (January-March), it was higher. The flow was adequate to provide water for up to 1000 residents. In 1997, Kenneth Wright's team conducted a hydrological analysis that estimated the amount of water from the primary source to be linked to the volume of rainfall, with a delay of approximately four months. This allowed for a steady supply of water to the citadel during the dry season (April to August) (Wright and Valencia 2009).

To harness the power of the spring, the Incas constructed sturdy walls using permeable stones measuring 1.4 m in height and 14.6 m in length. This method allowed them to filter the water from the main spring into a rectangular stone pit, which was then connected to a 749 m channel with a 3% slope. This system enabled the supply of water to the 16 springs located in the urban area of Machu Picchu. A secondary spring was also utilized, which flows into the collection channel approximately 80 m west of the main spring (Wright et al. 1997).

In another aspect, to reduce erosion in urban buildings, thatched roofs were used along with 129 surface drainage channels. These channels primarily flowed into the main drainage channel, which separated the agricultural and urban areas. Taking into consideration the water requirements of the cultivation terraces (platforms), it was estimated that they were primarily supplied by rainfall. However, as a backup system, water from the river located 500 m below was also utilized (Wright et al. 1999).

In 1994, Wright noted that he was pleasantly surprised to discover that even after four and a half centuries, the spring was still providing water to the citadel of Machu Picchu. He also observed that the drainage system was still operational (Wright and Wright 1996).

5.4.3 Amuna System

The term originates from Quechua or Aymara and has two meanings. In Quechua, it is "yacu hamunchu," which translates to "come the water." In Aymara, it refers to a person with a mute mouth, either due to performing ablutions or not having started speaking yet (De la Torre 2017).

The amunas are surface recharge systems located at the margins of riverbeds. Their objective is to capture water from streams in the mountains and transport it to desert areas or areas with low agricultural production through canals (De la Torre 2017).

Fig. 5.2 Mooring system. *Source* An ancestral system "sows" water in the high mountains of Peru (Samon 2021)

Similarly, the amunas are a traditional method of artificially replenishing aquifers to increase the water volume in springs during rainy seasons. These structures are typically located at the source of micro-watersheds (Apaza, n.d.).

This system functions by filtering water through irrigation ditches that run along the slopes of the mountains. The filtration process occurs due to the ditches being made of earth, without any layer of sand or cement coating. These ditches can extend for several kilometers and have low slopes, which allow the water to flow at a low speed as it passes through the ditch. It should be noted that the peasant communities are currently responsible for cleaning and maintenance (De la Torre 2017) (Fig. 5.2).

5.4.4 Irrigation Systems

The abundance of water resources ensures a consistent supply of this vital resource. However, the topography of the land, including slopes, had to be managed in order to fully utilize the agricultural resources. Engineering works on riverbeds did not begin with the Incas, but they were the ones who attempted to regulate the flow of water through irrigation systems and topographical modifications. This is the case with the creation of lagoons and reservoirs to provide water for agricultural canals, as well as the utilization of dried wetlands for agricultural activities. According to Beltrán (2013), this led to the development of a city with both infrastructure and settlements.

The agricultural and livestock production of our ancestors required the development of techniques to control the climatic and geographical adversities of the Andes. They mastered two natural resources, land and water, and thus, hydraulic solutions were born (Peralta 2013). As a result, pre-Inca communities began developing interventions in their watersheds to increase the availability of water during the dry season in response to the variable natural flow regime as early as 600 AD (Ochoa-Tocachi et al. 2019, p. 6).

Irrigation works became increasingly important, particularly for the coastal region, as people had to contend with water scarcity during the 560 s A.D. According to Ribeiro (2013), "thanks to the remarkable construction of canals, it was possible to transform the desert regions of the coast into arable land, to acquire soil for cultivation in the highlands, and to protect crops against drought." Consequently, the Puquios were created as underground aqueducts where water flows and discharges into the irrigation canals known as "cochas." Along the galleries, there are well-known "ojos"—infrastructures that facilitate the cleaning of the galleries. Other important infrastructure includes irrigation ditches and canals. The aqueducts were thousands of kilometers long and used to carry water to the fields. The flow of water was regulated by gates made of stone slabs (Alcayhuaman 2007).

The irrigation systems along the coast of Tahuantinsuyo have been preserved to this day. Among them are the canal system in the Tumbes River, which modern channels still follow the indigenous traces. Another notable system is the large diversion channel on the right bank of the Chira River, originating from Pazul and known as the buttress of the Cordillera. This channel leads to the estate of Solana and extends up to Chocán, covering a distance of 160 km. At the headwaters of the La Leche River lies Incahuasi, a significant population center of the Tahuantinsuyo civilization. The land in this area was irrigated by an irrigation channel that originated from the Yanahuanga lagoon. It is worth mentioning that this channel is currently abandoned. Another notable irrigation system is the Racarum Inca canal, located in the Lambayeque river basin, also known as the Chancay river. This canal, with its source in Cerro Blanco, irrigated the districts of Tucume and Mochumi. Additionally, the Chan Chan irrigation system utilized the waters of the Moche river to irrigate the lands of Chan Chan (Regal 1945).

From the engineering point of view, the Incas acquired knowledge through the influence of other cultures, both in technique and in the skill of stone work (Alcayhuaman 2007). Among these, we can mention the Moche culture, which greatly influenced the Incas in terms of infrastructure. The Incas preserved most of the canals built by the Moche people, and these canals still exist today with the names of their original builders, such as Sun Pisun and Chok-Chok (Leguía and Martínez 1914).

5.4.5 Production Systems

Peru and Bolivia are countries that boast a diverse array of micro-climates throughout their territories, resulting in a rich variety of flora and fauna species. The rugged and uneven terrain of the region creates various climatic challenges (Scherbosky 2013).

Therefore, innovative crop production systems were necessary to overcome these challenges. One of the main systems is the so-called Waru Waru.

The well-known camellones, also known as "Waru Waru" or "Suka Kullus" in some regions of the country, are raised agricultural beds constructed over ditches near water sources in terrain located between 3800 and 3900 m.a.s.l. The night temperatures in this region can vary between −5 and 0 °C. The dimensions of these structures vary from 4 to 10 m in width, 10 to 100 m in length, and 1 to 5 m in depth (Lhomme and Vacher 2003). The function of the Waru Waru was to control water and efficiently use flood lands. However, during times of drought and frost, they provide additional benefits such as surface or subsurface infiltration through the channels and radiative cooling at night, respectively (Kendall and Rodriguez 2015). The ditches also served the purpose of composting as algae grew in them. These algae were extracted and used as fertilizer for crops (Vásquez et al. 2016). The net radiation at altitudes above 4800 m is significantly higher than at lower altitudes. This, combined with lower air density, allows the water stored in Waru Waru systems to act as thermal regulators, enabling high agricultural productivity (Durand 2015).

Currently, they continue to be vitally important for agriculture in various regions of the country. Such is the case in the department of Puno, where these irrigation systems are still in use today, covering an area of 82,000 hectares within a 30 km radius. Various public and private organizations have invested in the reinsertion and recovery of traditional infrastructures, as demonstrated by the "Interinstitutional Project for the Rehabilitation of Waru Waru in the Altiplano (PIWA)" in 1989, which focused on the cultivation of quinoa crops (Chávez and Llerena 2015). On the other hand, in terms of statistics, as reported by Altieri and Nicholls (2001), the implementation of Waru Waru in the Huatta district led to an annual potato crop yield of 14 tons per hectare, compared to the previous yield. Therefore, their benefit is evident (Fig. 5.3).

5.4.6 Aqueduct System

The aqueduct system used by our ancestors has several examples still in operation today, including those built by the Incas such as the aqueducts of Macchu Picchu, Ollantaytambo, and Tipón (Solís et al. 2019). On the other hand, aqueducts were present in the headwaters of the Cumbre-Mayo streams, in the Jequetepeque River valley. This infrastructure extends approximately 400 m and carries water from the Jancat, Cumbe, Concejo, and Frailones hills to the city of Cajamarca. Likewise, the large embankment located on the slopes of the Chicama River, known as the Mampuesto de Facalá, was used as an aqueduct to transfer water from distant bodies of water up to 35 km away to the barren lands. It was built with adobe bricks and stones (Regal 1945).

However, the predominant culture in this region is the Nazca culture, known for their innovative subway filtration system designed to capture water for the water table. This system allowed them to have a reliable source of water for their crops, including avocado, beans, and pallares. The Nazca people developed this system due

Fig. 5.3 Camellones or Waru Waru, Titicaca basin, department of Puno. *Source* Revaluation of the traditional agro ecosystem of sukaqollos and agricultural development in Puno—Peru (Canahua 2014, slide 20)

to the frequent droughts caused by the underground river that ran through the area during that time (Peralta 2013). The use of groundwater has been given in various parts of the territory of the Nasca valley ponds or puquios where water was stored to irrigate the surrounding land which is now known as filtering galleries, detailing that its construction was developed by pre-Inca cultures, such as the Nazca and pre-Nazca culture. These galleries or aqueducts have a length ranging from 200 to 1,500 m. The rectangular sections are 0.60 to 1.20 m wide and 0.60 to 0.80 m high (Regal 1945). The importance of the Nazca aqueducts was highlighted in 1996 when an earthquake destroyed the public water network in the department of Ica. The population of Nazca had to rely on the aqueducts built by their ancestors to meet their basic needs and sustain agricultural production (Peralta 2013, p. 2).

In the aqueducts, there are helical section divers, also known as chimneys or breathers. Their main function is to maintain the infrastructure and regulate atmospheric pressure to ensure proper flow. These infrastructures can generate flow rates ranging from 13 to 67 L per second. Currently, there are a total of 29 aqueducts, of which only seven (7) are in optimal condition. For example, Ronald (2017) mentions Achato, Ocongalla, and Cantalloc as well as Orcona and Huachuca, which are known for their regular conditions (Fig. 5.4).

Fig. 5.4 Cantalloc aqueducts, district of Nasca and Vista Alegre, province of Nasca, department of Ica. *Source* Nasca Aqueducts (Castillo 2019, p. 15)

5.4.7 Storage Systems

The Incas built storage systems called cochas to manage water effectively. These systems involved creating a network to capture water from canals and aqueducts, allowing for storage and distribution. This enhanced water management for consumption by the population and for use in food production. According to Ancajima (2016), this system has the added benefit of improving soil fertility. The constant humidity in the soils of the oxbow lakes promotes abundant biomass production. Additionally, the system captures the erosion caused by rainfall, which would otherwise wash away nutrient-rich soil (Benavides et al. 2004).

The system of cochas or qochas, a Quechua term, is described as a collection of small natural water deposits that form concave depressions due to the accumulation of seasonal rainwater in the area. With the intervention of Inca engineering, these depressions were transformed into reservoirs of varying sizes, lined with stone or clay (Kendall and Rodriguez 2009). In addition, the Incas also constructed diversion canals that linked one oxbow lake to another and, subsequently, drained the water until it merged with a river or dissipated in the pampas (Benavides et al. 2004).

In general, the benefits obtained from this system are similar to another engineering system practiced by the Incas, known as waru waru or camellones. Both systems allow for the accumulation of water resources to be used during periods of scarce rainfall or to control the abundant flow of water, as needed. Also, the presence of water in these systems accumulates heat from solar radiation and mitigates the cold when frosts occur (Canziani, no date).

References

Alcayhuamán L (2007) Civil Engineering of the Incas. In: 5th Latin American and Caribbean conference for engineering and technology. http://laccei.org/LACCEI2007-Mexico/2007%20Copyright%20Pending/CEM218_AlcayhuamanA.pdf

Altieri A, Nicholls C (2001) Agroecology: principles and strategies for a sustainable agriculture in Latin America in the 21st century. http://agroeco.org/doc/pengue.htm

Ancajima R (2014) Tecnologías Ancestrales-sistemas Hidráulicos pre Incas e Incas. Ministerio del Ambiente, Lima. https://www.minam.gob.pe/diadiversidad/wpcontent/uploads/sites/63/2015/01/resumen1.pdf

Ancajima R (2016) Ancestral wisdom, a path in adaptation to climate change. XXVII Latin American Congress of hydraulics, Lima, Peru. http://investigacionesyproyectoshidraulicos.com/web/Material%20Cientifico/Articulos/Mas%20de%201000%20Articulos/826.pdf

Apaza D (n.d.) Las amunas: Ancestral technology of artificial recharge of aquifers in the Peruvian Andes. https://app.ingemmet.gob.pe/biblioteca/pdf/CPG14-266.pdf

Beltrán J (2013) Water and urban form in pre-Columbian America: the case of Cusco as a center of Inca power [Doctoral thesis, Polytechnic University of Catalonia]. http://hdl.handle.net/2117/95148

Benavides M, Inbar M, LLerena C (2004) Conservation and abandonment of terraces. Universidad Nacional Agraria La Molina, University of Haifa. http://www.lamolina.edu.pe/facultad/forestales/web2007/PublicacionesYRevistas/pdf/contenido.pdf

Canahua A (2014) Revaluation of the traditional agro ecosystem of sukaqollos and agricultural development in Puno—Peru. Ministry of Environment. https://www.minam.gob.pe/wp-content/uploads/2014/02/04-Alipio-Canahua-Humedales-Altoandinos-y-Agricultura.pdf

Canziani J (n.d.) Cultural landscapes and territorial development in the Andes. Pontificia Universidad Católica del Perú, Cuadernos de Arquitectura y Ciudad, pp 61–120. https://repositorio.pucp.edu.pe/index/bitstream/handle/123456789/28683/cuaderno_05_1.pdf?sequence=2&isAllowed=y

Carlotto V, Cárdenas J, Fidel L (2009) The geology, geomorphological evolution and external geodynamics of the Inca city of Machu Picchu, Cusco-Peru. Revista de la Asociación Geológica Argentina 65(4):725–747. https://revista.geologica.org.ar/raga/article/view/878/903

Castillo LJ (2019) Aqueducts of Nasca. Submission form: Lista Indicativa. Ministry of Culture. https://patrimoniomundial.cultura.pe/sites/default/files/li/pdf/1.%20Acueductos%20de%20Nasca%20-%20Esp_reduce.pdf

Chávez L, Llerena C (2015) Inventory of traditional and modern agricultural technologies for adaptation to climate change in the Andean zone of Peru [Degree thesis, Universidad Nacional Agraria La Molina]. http://repositorio.lamolina.edu.pe/handle/UNALM/2173

Cruz MA (2018) Andean cosmovision and interculturality: a look at sustainable development from sumak kawsay. Chakiñan J Soc Sci Humanit 5:119–132. https://doi.org/10.37135/chk.002.05.08

Durand D (2015) Some ancestral legacies, geomorphology in Peru and its contributions to land use planning tasks. Investigaciones Sociales 19(34):53–62. https://doi.org/10.15381/is.v19i34.11752

Kendall A, Rodríguez A (2009) Ancient agricultural infrastructure and its sustainability in the highlands and southern altiplano. Development and perspectives of the andenería systems of the central Andes of Peru. Cuzco, Institut français d'études andines. https://doi.org/10.4000/books.ifea.6118

Leguía, Martínez (1914) Geographic Dictionary of Piura. vol 1. Lima

Lhomme J, Vacher J (2003) Frost mitigation in the camellones of the Andean altiplano. Bulletin De l'institut Francais D'études Andines 32(2):377–399. https://doi.org/10.4000/bifea.6556

Llosa J (2008) Water harvesting in the Nepeña Basin. In: Elaboration and implementation of a national program for adaptation to climate change, with emphasis on selected areas of the selected zones of the Central and Southern highlands of the country. Technical Report. Lima, Peru,

CONCYTEC, pp 6–60. https://drive.google.com/open?id=0B0wk3XVEuBLkOV9yUUZnbGdPRHM
Miño L (1994) El manejo del espacio en el imperio Inca. FLACSO Sede Ecuador. https://biblio.flacsoandes.edu.ec/catalog/resGet.php?resId=44224
Monteverde L (2016) The ushnus and the cult of gods with attributes over water for the benefit of agricultural production within the Inca state administration. J Amer Archaeol 25:247–276. https://www.jstor.org/stable/27768520
Ochoa-Tocachi B, Bardales J, Pérez K, Acosta L, Mao F, Zulkafli Z, Gil-Ríos J, Angulo O, Grainger S, Gammie G, De Bièvre B, Buytaert W (2019) Potential contributions of pre-Incan water infiltration infrastructure to water security in the Andes. https://www.researchgate.net/profile/Boris-Ochoa-Tocachi/publication/349042071_Potential_contributions_of_the_pre-Inca_water_infiltration_infrastructure_for_water_security_in_the_Andes/links/601c2ce6a6fdcc37a8023d36/Contributions-water-infiltration-preinca-water-infiltration-infrastructure-potentials-for-hydro-security-in-the-Andes.pdf
Peralta P (2013) El agua en el Perú: pasado, presente y futuro. https://www.produccion-animal.com.ar/agua_cono_sur_de_america/63-Agua_incas.pdf
Ponce-Vega LA (2015) Puquios, qanats and springs: water management in ancient Peru. Agricul Soc Developm 12(3):279–296. http://www.scielo.org.mx/scielo.php?pid=S1870-54722015000300002&script=sci_arttext
Regal A (1945) Hydraulic policy of the Inca empire. Revista de la Universidad Católica 13:2–3. https://repositorio.pucp.edu.pe/index/handle/123456789/53531. https://repositorio.pucp.edu.pe/index/handle/123456789/53531
Revenga García JD (2017) The political organization and administration of the Inca empire. https://uvadoc.uva.es/handle/10324/25497
Ribeiro L (2013) Water in Inca and pre-Inca civilizations: myth, science and technology. VIII Iberian Congress on Water Management and Planning. http://revistas.lis.ulusiada.pt/index.php/8cigpa/article/view/351/pdf_34.
Ronald A (2017) Pre-Hispanic aqueducts...that still work. Inca Hydraulics. https://hidraulicainca.com/ica/acueductos-de-nazca/acueductos-prehispanicos-que-todavia-funcionan/
Samon C (2021) An ancestral system "sows" water in the high mountains of Peru. The daily Agora of water. https://www.elagoradiario.com/agua/sistema-recogida-aguas-amunas-peru-montanas/
Scherbosky R, Furlani N, González C, Carmona A (2013) NEW PARADIGMS—ANDEAN VISION OF WATER. XLIV AAEA Annual Meeting—San Juan. https://inta.gob.ar/sites/default/files/script-tmp-inta-ipafcuyo_nuevos_paradigmas-principios_andinos_de.pdf
Sherbondy J (1987) Hydraulic organization and power in Inca Cusco. Revista Española de Antropología Americana 117–153. https://revistas.ucm.es/index.php/REAA/article/view/REAA8787110117A
Solís Y, Mora D, Garbanzo I (2019) Guayabo National monument: water quality of the pre-columbian aqueduct. Tecnologia en Marcha 32:57–71. https://doi.org/10.18845/tm.v32i10.4881
De la Torre J (2017) Las amunas de Huarochirí: Agua, Innovación y Desarrollo. https://www.researchgate.net/publication/279177198_Las_Amunas_de_Huarochiri_Agua_Innovacion_y_Desarrollo
Varilla O (2015) Prehispanic dams in the Nepeña basin. Pre-Columbian water storage and infiltration infrastructures. Ancash, Peru. Knowledge management and capacity development to promote integrated water management in family agriculture (GIAAF). http://giaaf.pe.iica.int/getattachment/91b5b882-6557-4fa5-977f-c36103ad969d/1-4-Represas-Prehispanicas-de-la-cuenca-del-Nepena.aspx
Vásquez A, Mejía A, Faustino J, Terán R, Vásquez I, Díaz J, Vásquez C, Castro A, Tapia M, Alcántara J (2016) MANEJO Y GESTIÓN DE CUENCAS HIDROGRÁFICAS. Lima, Fondo Editorial-UNALM. http://www.untumbes.edu.pe/vcs/biblioteca/document/varioslibros/1209.%20Manejo%20y%20gestión%20de%20cuencas%20hidrográficas.pdf
Wright R, Valencia A (2009) Machu Picchu: Wonder of Civil Engineering. Lima, UNI

Wright K, Wright R (1996) Machu Picchu: its engineering infrastructure. USCID Water Management Conference. https://mountainscholar.org/bitstream/handle/10217/46816/101_1996-USCID-LasVegas.pdf?sequence=1#page=459

Wright KR, Kelly JM, Zegarra AV (1997) Machu Picchu: ancient hydraulic engineering. J Hydraulic Eng 123(10):838–843. https://ascelibrary.org/doi/10.1061/%28ASCE%290733-9429%281997%29123%3A10%28838%29

Wright KR, Zegarra AV, Lorah WL (1999) Ancient Machu Picchu drainage engineering. J Irrigation and Drainage Eng 125(6):360–369. https://ascelibrary.org/doi/abs/10.1061/(ASCE)0733-9437(1999)125:6(360)

Wright KR, Wright R, Zegarra AV, McEwan G (2011) Moray: Inca engineering mystery. American Society of Civil Engineers. https://ascelibrary.org/doi/abs/10.1061/9780784410790

Wright KR (2011) Mysterious Moray. Incan embellishment reflected upon the ground. Root J. Graduate students. Denver School of Architecture and Planning. University of Colorado, pp 59–65. http://wrightpaleo.com/wordpress/wp-content/uploads/2011/11/Moray-ROOT.pdf

Chapter 6
Rural Technologies and Practices in Ancient Peru

Ney Rios

Abstract This chapter will address two of the most representative technological works related to water management in Peru. The Cumbemayo Canal and the AMUNAS. The first one is a pre-Columbian hydraulic work located in the tropical Andes of Peru. Built largely on rock, its design and construction demonstrate a high level of knowledge in hydraulic engineering. The canal was designed to capture water from springs, reduce water velocity, and mitigate erosive effects. This significant hydraulic project, located over 3500 m.a.s.l., had the primary objective of ensuring a consistent water supply to communities and cultivated fields, even during periods of low water or prolonged droughts. Due to its magical and religious significance, this hydraulic complex is recognized as one of the most important pre-Inca ceremonial sites in South America. It is still in operation today. On the other hand, the pre-Hispanic water management system was based on recharging and artificially using aquifers, which involved planting and harvesting water. This system utilized hydraulic technology to retain rainwater in the soil and required active community participation for management and maintenance. It was framed within an Andean cosmovision to ensure a constant water supply to the communities. The Amunas are an infiltration system that considers the natural and climatic elements of a region to facilitate groundwater recharge in high areas and provide water for human consumption or agricultural production in low-lying areas.

Keywords Water irrigation · Andean technology · Rural constructions · Ancestral traditions

N. Rios (✉)
CATIE, Turrialba, Costa Rica
e-mail: ney.rios@catie.ac.cr

G. A. Corzo Perez et al. (eds.), *Aqua Vitae*, The Latin American Studies Book Series,
https://doi.org/10.1007/978-3-031-57731-4_6

6.1 Location

The Cumbemayo canal is situated at an altitude of 3600 m.a.s.l. in the tropical Andes of Peru. It is located 9 km southeast of the city of Cajamarca, in the province of Magdalena, which is part of the Department of Cajamarca in northern Peru. One of its main sections has coordinates of −7.191263° South Latitude and −78.581413° West Longitude.

6.2 The Cumbemayo Canal Objective

The canal, which spans 9 km in length, was constructed to collect rainwater from the western side of the Andes (Pacific slope) and redirect it to the Atlantic slope for various purposes.

6.3 Description of the Technology (Characteristics)

Built approximately 1500 years ago by the Caxamarca Culture, the site known as "Cumbe Mayo" derives its name from the Quechua language. "Kumpi Mayu" translates to "well-built water channel," while "Humpi Mayo" means "narrow river." Many sections of the canal, measuring 35–50 cm wide and 30–65 cm deep, were carved into living rock. These sections feature flow regulators designed in a zigzag pattern to reduce water speed and erosion, spillways, and numerous petroglyphs depicting the cosmogony of the ancient Caxamarcas, who built the canal.

According to Deza and Castañeda (2020), there are certain characteristics that demonstrate the mastery of hydraulic engineering knowledge over three thousand years ago, such as:

- The structure is divided into three distinct sections, each serving a specific purpose—either sacred or productive. These sections were constructed using varying techniques and at different points in time.

- First Section: Major Ritual Section

Carved entirely from rock, this initial 853 m section is carved on a trachyte mantle and was constructed with a religious purpose. In this section, water flowing from the Pacific slope to the Atlantic slope is diverted and utilized, potentially symbolizing the connection between the two hemispheres of the brain or the integration of thought and emotions.

The beginning of this section of the canal is a complex hydraulic system that includes an intake (or catchment) located at an altitude of 3578 m.a.s.l., where four streams converge: El Altar and a solid platform with drawings carved on the surface of the rock. In this area, two pairs of feet are drawn with directions pointing toward the south and north. Additionally, there is infrastructure in place to monitor the inflows, which is noteworthy.

- Second Section: Agricultural Rituals

This second section, spanning 2500 m, runs through agricultural areas that are also carved into the mountainside. In certain areas, artificial borders have been constructed using stones and filler.

- Third Section: Final Section

This 5650 m section, constructed on natural terrain and extending into the city, was designed to transport water for practical irrigation purposes and to provide water to two large reservoirs during periods of low water.

- Technical specifications:
  - Water flow control is achieved through the use of geometric figures, such as circles and squares, in segmented channels.
  - Control of water velocity: The canal has been constructed with an almost flush slope, with a gradient of one and a half millimeters per meter of length. This design allows for low velocities and ensures that there are no significant variations in its sections. It also includes narrow, wide, and zigzagged areas designed to decrease water velocity and dissipate energy.
  - The first sections of the canal are equipped with gates, including a water meter.
  - Elevated aqueducts can be used to cross depressions in the terrain.
  - A 10-m tunnel is required to cross a section of the mountain.
  - The presence of ritual zones and petroglyphs is linked to the onset of rainfall, agricultural activities, and an astronomical observatory.

Pictures

6.4 AMUNAS Objective

Establish a system to capture runoff water for subway water recharge to conserve water in springs and streams for dry periods or low water levels.

6.5 Description of the Technology (FEATURES)

The Amunas system, derived from the Quechua word for "retention," is an ancestral technology that focuses on the artificial or human recharge of aquifers in high Andean basins through natural infiltration processes. This system takes into account the unique natural conditions of the territory and is designed to promote sustainable water management practices. Some recent research, such as Alescastre (2012) and Ochoa-Tocachi et al. (2019), has identified the following subsystems and characteristics in the Amunas.

The hydraulic subsystem consists of a network of canals and irrigation ditches designed to collect runoff water from rainfall, springs, or diverted streams in the upper regions of the high Andean basins (above 3000 m.a.s.l.). The water is then retained in the soil through infiltration processes. The elements of this subsystem are (Fig. 6.1):

(a) Routing channels refer to the infrastructure that is constructed to capture and direct the flow of runoff water toward specific areas of interest or infiltration. There are two types of channels: long channels that allow water to be directed to infiltration channels or infiltration slopes, and short channels that direct water to other watercourses or natural reservoirs.

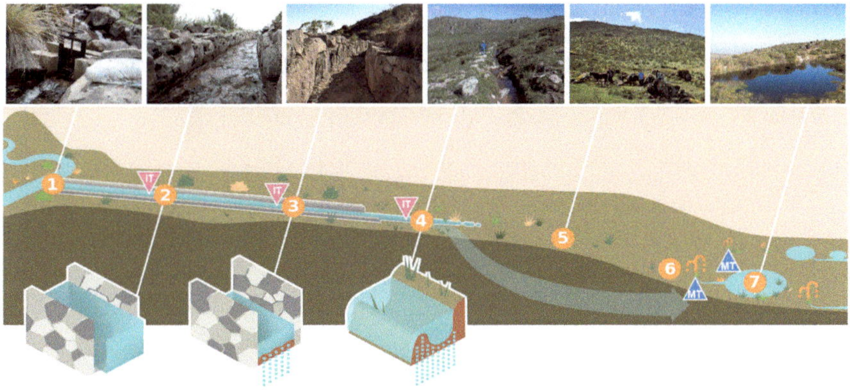

Fig. 6.1 Conceptual model of the hydraulic subsystem of the Amunas. The figure shows 1–2 routing channels, 3–4 infiltration channels, 5 infiltration slopes, 6 springs, and 7 Balsas. This information was sourced from Ochoa-Tocachi et al. (2019)

(b) Infiltration canals are earthen canals with minimal slopes designed to transport water to infiltration slopes while also allowing water to infiltrate along their path.
(c) Infiltration slopes are rocky or crevassed areas with high infiltration capacity that receive water from the infiltration channels and incorporate it into the subsoil. They serve as natural reservoirs or water recharge areas, allowing water to be stored in the soil and slowly released over time.
(d) Springs, also known as water springs, form naturally in areas where water is supplied through infiltration processes on hillsides or infiltration canals.
(e) Balsas are small reservoirs or bodies of surface water that are used to store or regulate the flow of water through infiltration. They had two purposes: on the one hand, to store water for direct use, and on the other hand, to enhance the groundwater infiltration process.

The community participation subsystem is of great importance as it is linked to activities related to the establishment, maintenance, and management of the Amunas. The motto "who does not work, does not have water" emphasizes the significance of community involvement in ensuring access to water. For the communities, the Amunas serve as a symbol of life and are considered sacred places. As such, the development of sacred rituals is an integral part of their operation and permanence. Several studies indicate that the amunas would not be possible without community organization and participation.

References

Alencastre Calderón A (2012) Alderón. Las amunas: siembra y cosecha del agua. Revista agroecológica LEISA 28(1):36
Deza J y Castañeda H (2020) Hierofanía y matemática del canal Cumbemayo, Cajamarca, Perú. Revista Ciencia y Desarrollo 23(4):67–101
Ochoa-Tocachi BF, Bardales JD, Antiporta J, Pérez K, Acosta L, Mao F, Buytaert W (2019) Potential contributions of pre-Inca infiltration infrastructure to Andean water security. Nat Sustain 2(7):584–593. https://doi.org/10.1038/s41893-019-0307-1

Chapter 7
Perspective on Risk Scenarios: Worldview, Perceptions, and Management of Water Resources in Mexico

Juan Cervantes Pérez and Alejandra Pacheco Mamone

Abstract This text is part of a multidisciplinary approach that integrates reflections from the fields of cognitive anthropology, climatic sciences, and ethnohistory. Initially, we will address how the Mesoamerican horizon has influenced the cosmovision, perceptions, and representations of water resources in present-day populations. We will discuss how this ontological framework and its transformations have influenced the management of water resources through processes of appropriation, negotiation, and conflict. This will exemplify diverse profiles of water management, taking an integrated approach to vulnerability and risk that is modulated by ethnic and regional aspects of history, landscape, and climate. Finally, we will discuss the challenges typically encountered by indigenous communities in situations involving environmental impact, as well as the issues of governmental institutional fragmentation that arise in the context of resource appropriation and management.

Keywords Worldview · Water management · Risk · Vulnerability

7.1 Reflections on the Cosmovision of Water, Historical, and Current Aspects

In Mexico, many indigenous peoples and communities inhabit territories with abundant precipitation and surface water, which contribute to the country's streams and aquifers. These communities also play a crucial role in preserving vegetation cover

J. C. Pérez (✉)
Licenciatura en Ciencias Atmosféricas, Universidad Veracruzana, Veracruz, Mexico
e-mail: jcervantes@uv.mx

A. P. Mamone
Community Development of the Tuxtlas, A.C., INAH National Institute of Anthropology and History, Tuxtla Gutiérrez, Mexico
e-mail: alinapacheco2024@hotmail.com

G. A. Corzo Perez et al. (eds.), *Aqua Vitae*, The Latin American Studies Book Series,
https://doi.org/10.1007/978-3-031-57731-4_7

in the upper regions of watersheds, as noted by Peña et al. (2010) and Villagómez and Gómez (2020).

Melville and Cirelli (2000) note that until recently, there was a prevailing perception that there was an abundance of water resources on the planet. It is common for the issue of water scarcity and crisis to be oversimplified as a matter of physical water scarcity due to population growth and the increasing social and economic demand for water resources for various purposes. However, these authors argue that this crisis is also a result of the degradation of surface and groundwater sources. Capital investments and technological innovations alone are not a sufficient response to the current problem. Thus, although the water crisis has many facets, there are three dimensions that cannot be left unaddressed: cultural, ecological, and political.

In terms of culture, the following characteristics are noteworthy:

Contemporary indigenous cultures have a pre-Hispanic past, which has undergone changes, transformations, and resignifications in their own values and practices. Cultural and religious syncretism has played an essential role, because rather than remaining pure, cultures have integrated pre-Hispanic components and the Catholic religion that arrived with the conquest. The cult of water and agricultural rituals are ceremonies that have transcended time and historical periods, which still retain much of their pre-Hispanic origin. Currently, we are facing a series of transformations, adaptations and changes that respond to their social, cultural, and economic contexts. For this reason, indigenous cultural manifestations are not conceived as the direct and uninterrupted continuity of the pre-Hispanic past but are visualized in a creative process of constant re-elaboration, sustained at the same time in very remote roots (Broda 2001a).

There are various origin and creation myths in different cultures that aim to explain their origins. Water often plays a significant role in these myths. Sometimes, it is associated with destruction, while at other times, it is linked to the creation of the world and natural resources, as well as domesticated crops such as corn and aquaculture fauna. Rituals play a significant role in many cultures. They can be categorized into annual, life cycle, or rebellion rituals. Some examples of rituals include agricultural and rain petition ceremonies, as well as healing ceremonies where water is a vital element. In the vast realm of agricultural rituals and traditional methods of maintaining harmony with nature, which are still observed by the majority of indigenous communities in Mexico, crucial aspects of the pre-Hispanic worldview and its fusion with the Catholic faith have been conserved (Broda 2001b).

Although piped drinking water systems have been introduced in many indigenous populations, a significant portion of them still rely on domestic wells and untreated spring water for their survival (Villagómez et al. 2013). Therefore, there are several ecological issues related to indigenous peoples, including:

An important part of the indigenous localities are settled in the upper parts of the watersheds, corresponding to forested areas, with an essential hydrological function for the recharge of aquifers and water supply for cities and irrigation areas (Martínez 2017).

In other words, we have the diversity of indigenous peoples with their own languages, social organization, and culture, in contact with ecosystems and important levels of biodiversity. These peoples have a reservoir of ecological knowledge and vernacular technologies that are applied to the management of natural resources as part of their natural and cultural heritage, which is the product of a hydraulic culture that dates back to the application and use of an ancestral technology, as research on the subject has shown (Villagómez and Gómez 2020).

For the analysis of the relationship between indigenous peoples and water, we consider three areas of indigenous cultures that involve forms of interaction, symbolization, appropriation, management, and use of water resources: (a) Cosmovision and forms of symbolic relationship with the resource, (b) conflicts and synergies around water management, and (c) risk scenarios and public policies in water resource management.

In the first part of our reflective approach, we aim to understand the multidimensional axes that make up and articulate the worldviews surrounding water resources in different conditions such as river currents, lake water, sea water, groundwater, and rainwater. We will observe how continuities can occur and how cognitive constructs that support particular views are re-elaborated by different ethnic groups, with a focus on the State of Veracruz.

According to Broda (2016), the ancient Mesoamericans had a well-structured worldview that integrated their beliefs about the environment they inhabited and the cosmos in which they believed human life existed (Broda 2016, p. 3). Recently, other researchers, such as Lorente Fernández (2011) and Martínez Villaroya (2019), have further developed the concept and emphasized the importance of not only observing continuities but also disruptions and adjustments that deviate from the static nature of these processes. On the other hand, to comprehend cognitive interrelationships, it is essential to consider how the hydrological cycle is perceived. This includes the interconnected flows of surface water, groundwater, seas, and coasts, as well as precipitation such as rain and hail, and their interactions with winds, hurricanes, and storms. Broda highlights that in Mesoamerican cosmology, hills and mountains were viewed as containers or vessels filled with water. They were believed to have their own guardians and protective entities (Broda 2016, p. 14). The subterranean waters converge toward a vast space beneath the earth known as Tlalocan, the paradise of the rain god. The same principles have continued to the present day with recent adjustments and incorporations (López Austin 2012; Trejo Barrientos et al. 2013). Chantal Huckert (2020) mentions that the House of the Quetzal or Green Feathers, known as Quetzalcallan, was believed to be the dwelling place of Tlaloc, the deity of rain and storms. The space represented a border because the struggles and conquests were pushing the borders of the Triple Alliance empire toward more distant territories (Huckert 2020, p. 322). Venter (2020) also notes that in various regions of Mesoamerica, including Nahua communities in northern Veracruz, hills hold significant cultural and spiritual value as "sacred mountains," "mountains of sustenance," or "mountains of water." It is believed that they are hollow and serve as liminal points connecting the underworld, terrestrial world, and the firmament. Markens, on his part, believes that the concept of a serpent serving as a guardian of the hill, its water, and natural resources is prevalent (Markens 2013, p. 73).

Let us now consider some aspects that help us understand the complexity of the line connecting past and present stages, revealing tensions and adjustments in various regions and historical periods.

Lorente (2012, 2017) focused on the complex processes that link water cycles and the forms they take in current worldviews. He proposes a concept called the "cosmic razzia," which could be defined as a large atmospheric process of extracting essences. This complex process articulates a cosmogonic cycle of exchange and reciprocity, integrated by two figures and governed by two axes. On the one hand, the figure of the ahuaques (from the Nahuatl word for owners of water), rain spirits that inhabit the springs and the tesifteros, graniceros or atmospheric ritualists who "understand the weather" and have the gift of interceding between these spirits and the human community. For the southern regions of Veracruz, the names will change, but they will still refer to the same phenomenon known as sun rays or sparkles. Lorente explains that in this dialectical process, the evolution of terrestrial life is interconnected with the life of the underworld, while the aquatic life of the ahuaques sustains life on earth. One represents the cosmic cycle, while the other represents the fertilizing contradiction. Both elements are integrated into the concept of "time" (Lorente 2006, 2008, 2009, p. 98, 2017).

It is important to note that there is an associated ritualistic system around these symbolic systems, with specialists linked to various processes, and broad community participation. Martínez and Van't Hooft (2012) addressed the petition for rain in Ichcacuatitla, Chicontepec, Veracruz, noting that the ritual extended over several days. During this time, preparations and rituals were carried out, including offerings to rain entities such as Atlatlacualtiliztli and Apanchaneh, the owner of water, as well as other guardians. In his study conducted in Coutla, located in the northern highlands of Veracruz, Romero Vivas (2003) highlights the perception of the world as a sphere surrounded by water, with interconnected vessels between different levels. There are forces represented by the "owners" of rivers and lakes that are in constant interaction with humans. Tensions and conflicts in an eternal balance are followed by acts of reciprocity between deities and humans (Romero Vivas 2003, pp. 50–51).

With a different approach, Ochoa Ávila and Arias (2016, p. 63) conducted studies in the Huasteca Veracruz area to explore the Teenek worldview. They observed that changes, conflicts, and resistance are reflected in the oral traditions of these communities. The Teenek people of Veracruz, similar to those in San Luis Potosí, refer to the myth of Thunder in their oral tradition. The Teenek people of San Luis Potosí refer to thunder as Maamlaab, while the Teenek people of Tantoyuca of Veracruz call it Junkil aab, which translates to "the one who accompanies the rain." An example of the crucial role of water can be seen in the Teenek communities of San Luis Potosi. These communities are primarily situated in the foothills of the Sierra Madre Oriental, where they experience storms, hurricanes, and rainy northes every year. These weather events bring moisture from the Gulf of Mexico to the mainland, resulting in an average annual rainfall of 2800 mm, and sometimes even more. Perhaps that is why the world is symbolized as a rhombus: "square, but elongated, with four cardinal points," one in each corner. Like a ball floating in a bucket of water, everything is surrounded by water and has clear boundaries that can be dangerous to cross. The meaning of water is ambiguous. Water originates and sustains the world, giving life to all living beings. Children are born surrounded by water, and it is also essential for growing crops, cooking, and drinking. But it can also be a threat, as scarcity or overpopulation can deny men access to it (Peña et al. 2010).

On the other hand, with regard to the methods of learning, initiation, and legitimization among ritual specialists[1] who are associated with various forms of water protection and rain spells, Pérez Apango (2020) emphasizes the significance of a tradition that places great importance on dreams. Ritual specialists in the Huasteca region mention that they learned about their teachings, behavior, and the relationship between thunder and lightning, as well as aquatic entities of rivers, streams, and lagoons through their dreams (Pérez Apango 2020, p. 85). Another important aspect that allows us to observe the theme of continuity and disruptions in the cosmogonic legacies linked to the theme of water and its transformations is provided by Martínez González (2006) when examining the function of the nahualli and its contemporary updates. Currently, Martínez González notes that in certain modern Nahua communities, such as those in the Huasteca, the Malinche region, and southern Veracruz, the belief in nanahualtin who can manipulate weather patterns and water cycles for the betterment of the community still persists. Martínez González has documented how this concept remains relevant and has been updated to the present day (Ibid 54). As Villagómez points out, every culture has a variety of origin and creation myths to explain its origins, and water often plays a significant role in these myths. Sometimes it is associated with destruction, while at other times it is linked to creation. Myths often address concrete problems related to water scarcity, with some alluding to the loss of this precious resource, while others celebrate its abundance (Villagómez 2013).

Let us now examine some examples from the southern region of Veracruz, specifically in the Los Tuxtlas area and the Santa Martha mountain range. Delving deeper into the southeastern region of Mexico, specifically within the state of Veracruz, lies the Sierra de Los Tuxtlas. This area boasts two volcanoes, San Martin Tuxtla and San Martin Pajapan, and is situated in a tropical region with immense potential for high biodiversity and cultural significance. The region is home to a Biosphere Reserve, as designated by CONANP-SEMARNAT in 2006, and has been studied extensively by Gutierrez Garcia and Ricker in 2011. Its archeological richness has also been extensively documented (Budar and Ladrón de Guevara 2020; Ciphers 2012). Extensive documentation exists on the relationship between Nahua and Popoluca groups in

[1] The term "ritual specialists" is commonly used in ethnohistorical and anthropological studies from various perspectives, including symbolic anthropology and cognitive anthropology. This category includes various subjects that involve knowledge of ritual practices associated with diverse cosmovisions. These practices can be used in different areas to perform various rituals, such as requesting rainfall, caring for agricultural practices during changing seasons, and utilizing diverse ethno-climatic knowledge. Some practices are linked to natural resources, while others are related to human health and disease processes. For instance, in the case of water cycle processes, tesifteros and graniceros can be cited as examples (Lorente Fernandez, D. 2017, Los graniceros de la sierra de Texcoco: repensando el don, la experiencia onírica y el parentesco. Ritual. "Dimensión Antropológica," Year 24, Vol. 70, May-August 2017).

this region and their environment. Their cosmovision reflects a deep interconnection with the natural resources they rely on, including mountains, volcanoes, and water from rivers, lagoons, and coasts. This has been extensively studied and recorded by scholars such as Lazos and Pare (2000), Durand Smith (2005), García de León (2011), and Vargas Montero (2010). In Nahua cosmology, the chaneques and their various hierarchies are small beings who act as guardians of river and stream courses. They can either favor humans or mislead them, causing losses depending on how they perceive human behavior (Lazos Chavero-Pare and Ouellet 2016; Pacheco 2013). In the region where the Popoluca people are predominant, there exist powerful figures known as the lords of the land and water. Additionally, there is a diverse range of entities, including the old woman Tchistchima, the Hunchitsu, an anthropomorphic being that keeps watch over the highest areas of the jungle, and the xonoti-lu'putji entities. These entities take the form of small felines or monstrous donkeys and oversee the moral relations among community members, as well as the reciprocal relationship with the natural resources of the land and aquatic fauna. (Blanco 1999; Cárdenas et al. 2005; García de León 2011; Oropeza Escobar 1998; Pacheco Mamone 2013; Pacheco Mamone 2018).

Another aspect we must consider in this reflection is the manner in which certain mythical figures are revived and conserved by ritual experts, like the ones we have previously mentioned. The figure of Agustín Agustín stands out, particularly for the Popolucan communities that are associated with rituals for the conservation of springs, rivers, streams, and the balance of rains and other hydrometeorological phenomena. Among the various hydrometeorological phenomena, the figure of Agustín Matehua or Payiwuyi (meaning wise old man or wise old tree) stands out. This figure is considered a wise man, a civilizing hero, and a guardian of traditions related to water and lightning. He represents an example of the dynamics and processes of adaptation (Báez-Jorge 2009, Aino 2016, Pacheco Mamone and Robles Guadarrama, 2014).

Thus, concluding this brief tour, we consider that the present-past links of a cosmogonic legacy are updated through the matrix of perceptions, representations, and adjustments in the cosmovisions of these groups, and we could consider it as a fluid mosaic, with nonlinear changes in response to both the processes of socioenvironmental impact and the dynamics of resilience building by these communities.

7.2 Water Management Conflict Scenarios and Synergies

In this section, we will examine the complexity involved in managing water resources, recognizing that it is closely intertwined with social differentiation and inequity. At the same time, let us remember that the most recent water conflicts occur in a context of accelerated transformation of modernizing neoliberal projects, and under a scenario of risks due to climate change impacts that require urgent attention. In the context of the 2030 water agenda, the report from the Ministry of Environment and Natural Resources (SEMARNAT) highlights the State of Veracruz, particularly the Pánuco area, the lower and middle Papaloapan basin, and the Coatzacoalcos region, as the most vulnerable regions to drought. (Semarnat, 2016, p. 9). De la Rosa et al. (2021, p. 187) used drought indices and a hydrological basin approach to show that 77% of the basins experienced a decrease in water during the period of 2014–2020. The inadequate management of solid and industrial waste contributes to the deterioration caused by contamination, and this process must be addressed. According to Domínguez Serrano and Castillo Pérez (2018), out of the 2517 drinking water and sanitation operating agencies established in 2009, only 48.2% provided services to both rural and urban areas, while 41.8% exclusively served urban localities. The authors note that the municipalization of drinking water and sanitation services occurred within a context of unequal territorial conditions (Domínguez Serrano and Castillo Pérez 2018, p. 480). In order to illustrate the point, we will start by providing historical examples of tensions and conflicts in Veracruz. Then, we will examine several instances of successful management and agreements.

Several researchers agree that while the resource may be abundant, its actual availability is strongly affected by deterioration, supply source affectation, contamination, geographic dispersion, and inadequate management processes (Paré 2009; Roca 2019). This can lead to social conflicts, where individuals employ various mechanisms to achieve their demands and fulfill their needs. According to the Conflict Observatory developed by IMTA in 2015, over 80 types of water-related conflicts have been recorded in the basins where indigenous communities are settled. Paré (2017) emphasizes the significance of basin approaches in resolving conflicts related to water resources. He argues that a society goes beyond its territorial boundaries and utilizes resources from non-contiguous territories, especially those from its supplying basins. Paré terms this concept as "social metabolism." Applying this concept to water issues helps in comprehending the balance or imbalance between consumption, production, and waste disposal, as well as the reintegration, absorption, and pollution of water resources (Paré 2017, p. 287). Ávila García (2016) argues that there is a process of coloniality in the management of water resources. She uses water as an illustrative example, as it is often managed under the logic of capital and is overexploited by industries such as export agriculture, extraction, transformation (oil, pulp and paper, chemical, automotive), and agroindustry (farms, biodiesel). (Avila Garcia 2016, p. 22).

The following are some examples of conflicts and their possible resolutions:

Roca (2019) examines the historical circumstances that resulted in intense conflicts in the Acayucan region. Since the 1960s, the primary source of water for human consumption has shifted from artesian wells to canalization and piping systems due to increasing demand for water supply. The researcher highlights this transformation. Since the 1980s, there have been ongoing demands and complaints about the situation of the Platanillo spring, which is located in Soteapan and feeds into the Michapan River. In addition to Acayucan, 18 mountainous communities rely on this resource for their water supply. In 2000, the aquifer produced approximately 500 L of water per second, which is equivalent to 1800 cubic meters per hour. As a result, Acayucan received a total of 540,000 L of water per hour, or 12,960,000

L per day. Twelve years later, the demand for water supply doubled during the dry season. Numerous conflicts, demonstrations, and protests involving the seizure of municipal buildings took place during the 1990s and 2000s. The author emphasizes, however, that it is important to recognize that in the midst of these conflicts, bonds of solidarity were also formed, bringing together diverse individuals to work toward a common goal. The solidarity of values has been evident in various mobilizations that have taken place in Soteapan, starting from 1989 when Acayucan was deprived of water for six days, to the most recent one in December 2016. "There has been a sense of community belonging" (Roca 2019, p. 319). Murillo Licea (2017) highlights that the Papaloapan basin can serve as a model for studying conflicts related to indigenous territorial boundaries. He points out that such violations not only have spatial implications but also symbolic ones, which unfortunately seem to be overlooked by government agencies. The construction of large dams in indigenous territories may have significant impacts on the affected communities. These works can lead to the fragmentation of settlement and interaction dynamics, as well as the erosion of cultural values (Murillo Licea 2017, p. 257). Similarly, in his historical study of Poza Rica and other regions of Veracruz, Kloster (2017) notes that the escalation of conflicts began to create a divide between 2008 and 2011. Over the years, the relationship between the communities in the Chicontepec region and Lagunilla, Papantla, with Pemex has resulted in drastic landscape transformations. These changes have also led to the emergence of new scenarios for crop economy, which have caused a range of negative effects on water resources (Kloster 2017, p. 192). In his analysis of the period 1990–2006 in the Córdoba-Orizaba area, Agüero Rodríguez highlights that socioeconomic demands often took precedence over socioenvironmental demands. This suggests that the perception of environmental values is still in its early stages and can be considered reactive. Frequently, socioenvironmental conflicts dissipate once the socioeconomic demand is met and the state intervenes (Agüero Rodríguez 2010, p. 220).

For their part, Paré and Robles have extensively documented conflicts in the Sierra de Santa Martha and Los Tuxtlas regions. Paré (2017) reflects on the issues present in the Texizapa-Huazuntlán micro-watershed. The area is situated in the upper basin of the Coatzacoalcos River and includes various communities within the municipalities of Mecayapan, Tatahuicapan de Juárez, and Soteapan. It is located between the San Martín Pajapan and Santa Martha volcanoes, within the Los Tuxtlas biosphere reserve. The municipality of Tatahuicapan is home to the Yuribia dam and a water treatment plant that collects and purifies water for 80% of the inhabitants of Coatzacoalcos, 100% of Minatitlán, 40% of Cosoleacaque, the company Tereftalatos Mexicanos de Cosoleacaque, and over 13,000 rural residents. From the beginning, the communities resisted and demanded social services in exchange for their involvement in the dam project, which aimed to bring water from the highlands to the cities. Government institutions refused, citing constitutional Article 27 which designates water as a common good (Paré 2017, p. 291). Since the dam was built in 1985, there have been tense relations and interactions among the inhabitants, government institutions, and political actors in the cities of Coatzacoalcos. The diverse perceptions and interests have led to cycles of conflict, negotiation, and further conflicts. This has resulted in valve closures and suspension of the resource flow to the coastal cities (Pare and Robles 2006: 87). In the challenging context of 2006, we witnessed the emergence of a new opportunity for progress, thanks to the collaboration between various peasant actors and a civil society organization known as the Intercommunity Watershed Committee (Moreno Arriba 2016, 2019; Robles Guadarrama and Pacheco Mamone 2013). This case, as well as others, demonstrates the potential that can arise from synergies and cooperation.

Vidriales (2016) analyzed the successful management process that was implemented in the Pixquiac River subbasin. This process is contextualized within the dynamics that supply water to the city of Xalapa. The author highlights that despite receiving above the national average rainfall of 1402 mm per year and an annual average of 760 mm, the city of Xalapa has been struggling to meet the increasing demand for water supply. However, it presents problems with water supply due to the water management and distribution model (Vidriales 2016, pp. 1, 13). A wide variety of stakeholders have been integrated into the co-management of the Pixquiac River basin. The experience aims to integrate research and action to address concrete problems in the subbasin and develop decision-making tools. Additionally, it involves the establishment of a participation platform that brings together social and governmental actors. The project has promoted the Pixquiac River Basin Committee as a participation platform, along with a financing mechanism to implement the actions outlined in the management plan. This plan includes the Compensation Program for Environmental Services of the Pixquiac River Basin, as described by Paré and Gerez (2011) and Vidriales (2016). In another regional context, but within the same search for synergies, are the co-management processes in the Texizapan Huazuntlán watershed, which we have mentioned. The co-management experience began in 2006 when four mountain communities came together to share their experiences in resource management for conservation and sustainable production. However, in 2007, a severe hydrometeorological event caused significant devastation in the region, which led to a new risk scenario. This event mobilized the communities to work together in a new phase of cooperation and synergy to find ways to recover from the impact. Within this framework, and with the cooperation of a civil society organization called Decotux, A.C., an intercommunity committee was created for the Texizapan Huazuntlan watershed in Veracruz. The committee initially consisted of six communities, but it has since grown to include thirty communities from the municipalities of Tatahuicapan de Juarez, Soteapan, Mecayapan, and Pajapan. Community authorities have also participated in the committee's formation. The active functioning of the committee, which is integrated by community authorities designated in community assemblies, and the technical interaction with the NGO, represented a new impetus. Interactions with government institutions have been renewed, along with various contacts and collaborations with several civil society organizations and academia (Moreno Arriba 2016, 2019; Pare and Robles 2008; Pacheco Mamone and Robles Guadarrama 2023; Robles Guadarrama and Pacheco Mamone 2013; Robles Guadarrama 2008). The committee was able to develop an integrated management plan at the watershed level, along with several applied technical models for ecological restoration and sustainable productive projects. Since 2017, it has been possible to implement activities contextualized in the processes of adapting to climate change through a work program that includes climate stations, water resource management, conservation, and sanitation strategies (Pacheco Mamone and Robles Guadarrama, 2014; Robles Guadarrama and Pacheco Mamone 2016; Pacheco et al. 2018).

7.3 Final Reflections on Risk Scenarios and Public Policies

In order to effectively formulate and implement public policies related to the management of a country's water resources, as well as to encourage participation in integrated resource management, attention is given to three levels of basins: (i) macrobasins, which correspond to large hydrological systems, (ii) subbasins or second order basins, and (iii) micro-basins, which are the smallest level of basins (Villagómez et al. 2013).

The dynamics of territorial organization among indigenous peoples, based on their agrarian structure, indicate that the level closest to their social reality are not the large regions, such as the watersheds defined by the National Water Commission for administrative purposes, but rather the micro-watersheds. These are delimited according to water resource management agreements among colonies, ejidos, communities, towns, and, at the most complex level, municipalities. (Villagómez and Gómez 2020).

Peña et al. (2010) argue that the establishment of indigenous territories involves both symbolic and practical appropriation, and that these two dimensions are intertwined and reflect the community's exercise of governance. As an example, the authors illustrate their experience with Tampaxal, a Tének community comprising nearly 6000 residents, of which 550 are ejidatarios. The community consists of 17 localities (barrios or sections) situated in rugged terrain. The 16 neighborhoods of Tampaxal must ensure a reliable supply of domestic water from sources they recognize as their own: the springs located in the caves and rocks of the mountainous wall that surrounds them. For this purpose, they utilize community organizations, which serve as a fourth level of government and are only partially connected to municipal authorities. For instance, they manage to include councilors from Tampaxal in the Aquismón City Council.

Pacheco Mamone and Robles Guadarrama (2023) refer to the ejido general assembly, neighborhood assembly, and water committees for the purpose of water supply.

The general neighborhood assembly, which convenes every two months, is the highest governing body in each neighborhood. It is responsible for making agreements and assigning responsibilities. Unlike the ejidal assembly, which is composed solely of ejidatarios, the neighborhood assembly includes all residents of the neighborhood who are over 18 years of age, regardless of gender or land ownership status. However, the opinions of married men are often given more weight, while women's participation is sometimes relegated to a secondary role.

The committee is the fundamental unit of the neighborhood organization. Sometimes, there is only one person in charge. Committees are responsible for addressing specific issues such as water management, schooling, and others. A barrio committee has jurisdiction over a delimited and relatively small group: the barrio. The obligation of the ejido committees is to supervise and coordinate the functions of the community committees. The water committee in each barrio is responsible for procuring and organizing the necessary equipment for the job, as well as supervising and maintaining the water infrastructure, including storage tanks, pipelines, and stopcocks. He is responsible for ensuring that there is a sufficient quantity of water during his shift and that it is not diverted by another committee.

Differences between neighbors in the same neighborhood are resolved by the water committee, which proposes certain solutions. If the neighbors continue to disagree, they can bring their opinions to the neighborhood assembly, which will ultimately make the final decision. In the event of a dispute between two neighborhoods or two water managers, the ejidal commissariat serves as an arbitrator. If the conflict is more significant, it is escalated to the ejidal assembly, which has the final say.

The person in charge of the water committee, along with the entire system of community charges, performs their functions without payment. They offer their work as a community service and cannot refuse the appointment without a valid justification presented to the neighborhood assembly.

The neighborhoods, whether or not they receive water from the general system, are supplied by other springs or wells located on their properties. In such cases, each neighborhood establishes the conditions for using the water and the resulting obligations.

Tampaxal's domestic water supply system combines a centralized system with individual sources that are decentralized by neighborhood or household. This allows the community to maintain its independence from the municipality in an issue as crucial as satisfying its domestic water needs. Contrary to the claims of some governmental agencies regarding water supply to rural communities, the residents of Tampaxal have always been self-sufficient in managing their own water sources and ensuring a reliable supply for their families. They do so under the conviction that they have the right to use and enjoy the water sources in their territory. It is not their intention to place themselves outside of federal law. However, their primary concern is not obtaining concessions or permits from the National Water Commission. Instead, they are focused on the internal organization of the ejido and the autonomous management of the tensions and conflicts that inevitably arise from the water supply.

While there are examples, such as the one described above, that demonstrate progress, the role of the state still requires significant attention and effort. In 2017, Avila conducted an analysis of the State's role in promoting a water and territorial management model that primarily benefits national and transnational private actors. This model has negative consequences for rural communities, particularly peasants and indigenous people who are marginalized. The state's strategy has relied on legal and institutional changes that result in the dispossession and privatization of previously communal resources, including water and land. Additionally, the government has prioritized economic projects and investments in regions crucial for capitalist globalization. However, there is a lack of social agreements at the national level regarding the implementation of these changes, as well as a lack of social participation in the design and execution of water policies. Furthermore, affected populations have limited access to information and consultation regarding water projects and works planned for their regions and watersheds. This results in a neoliberal approach to water management that intensifies social tensions and creates divisions between social actors and the state. There is a lack of national-level social agreements regarding the implementation of these changes, no social participation in the development and implementation of water policies, and no provision for access to information and consultation with the population affected by water projects and works in the regions and watersheds.

Notwithstanding the above, it is also true that mechanisms for citizen participation in decision-making related to water conservation and usage have been and are being sought after. For example, Martínez and Velasco (2014) argue that in participatory democracy, individuals and communities take the lead in determining the direction of change, while authorities serve as collaborators in bringing about that change. To the extent that we are knowledgeable about human rights and proficient in accessing information, participating in decision-making, and seeking justice, our impact on the events that occur in our communities will be more substantial.

References

Agüero Rodríguez JC (2010) Entre las demandas reivindicativas y ambientales Conflictos por el agua en la zona metropolitana Córdoba-Orizaba, Veracruz, 1990–2006. Series 3. Digital Library of Humanities, Universidad Veracruzana, p 321

Aino V (2016) Naturalización del concepto “religión” y colonialidad religiosa. El caso de los nuntajiykiwi (popolucas de la Sierra) de Santa Rosa Loma Larga (Veracruz). » Interdisciplina 6(16):31–60. http://dx.doi.org/10.22201/ceiich.24485705e.2018.16.65634

Avila P (2016) Towards a political ecology of water in Latin America. Rev Estudios Sociales 55:18–31. https://doi.org/10.7440/res55.2016.01

Báez-Jorge F (2009) Coordenadas etnológicas e históricas en el estudio de las cosmovisiones mesoamericanas. Contrapunto 4(11):44–64

Blanco JL (1999) La integración de los popolucas de Soteapan a la sociedad nacional: desarrollo, democracia y ecología.Tesis de Maestría. Universidad Iberoamericana, México

Broda J (2001a) The ethnography of the feast of the Holy Cross: a historical perspective. In: Broda J, Báez-Jorge F (Coords.) Cosmovisión, ritual e identidad de los pueblos indígenas de México. Fondo de Cultura Económica, Mexico, p 528

Broda J (2001b) Introduction. In: Broda J, Báez-Jorge F (Coords.) Cosmovisión, ritual e identidad de los pueblos indígenas de México. Fondo de Cultura Económica, Mexico, p 528

Broda J (2016) El agua en la cosmovisión de Mesoamérica. In: El agua en la cosmovisión de los pueblos indígenas de México. Ed. SEMARNAT, Mexico, p 168

Budar L; Ladrón de Guevara S (2020) Water use and representation on the Gulf Coast. Universidad Veracruzana-Instituto Literario de Veracruz, p 377

Cárdenas PJ, Córdoba Ortiz A, Ortiz Maciel D (2005) De piedra Labrada para la Sierra. Cuentos y mitos.Edit. CONACULTA-FONCA-Universidad Veracruzana., Xalapa, Veracruz, México

Cirelli C, Melville R (2000) The water crisis. Memoria 134:27–31

CONANP-SEMARNAT (Comisión Nacional de Áreas Protegidas y Naturales-Secretaria de Medio Ambiente y Recursos Naturales) (2006) Programa de Conservación y Manejo Reserva De La

Biosfera Los Tuxtlas. Comisión Nacional de Áreas Protegidas, Dirección General de Manejo para la Conservación, Dirección Regional Centro y Golfo, CONANP-SEMARNAT, Mexico City, Mexico

Cyphers A (2012) The beautiful theories and the terrible facts: controversies about the Lower Preclassic Olmecs. UNAM-IIA, pp 144

De la Rosa A, Valdés-Rodríguez OA, Villada-Canela M, Manson R, Murrieta-Galindo R (2021) Characterizing water security with a watershed approach: case study Veracruz, Mexico. Ingeniería Agua 25(3):187–203. https://doi.org/10.4995/Ia.2021.15221

Domínguez Serrano J, Castillos Pérez E (2018) Community water organizations in the state of Veracruz. Analysis in light of the Latin American experience. Estudios Demográficos Urbanos 33(2), 469–503. ISSN 0186-7210; E ISSN 2448-6515. https://doi.org/10.24201/edu.v33i2.1756

Durand Smith L (2005) Los mitos y la conservación ambiental. Líder, Revista Labor interdisciplinaria de desarrollo regional, N°13, pp 215–226

García de León A (2011) Tierra adentro, mar en fuera. El puerto de Veracruz y su litoral a Sotavento, 1519–1821. Edit. Fondo de Cultura Económica, México

Gómez Martínez A, Van't Hoof A (2012) Atlatlacualtiliztli: the rain petition in Ichcacuatitla, Chicontepec. Nahua language and culture of the Huasteca, pp 101–118

Huckert C (2020) Iconography of aquatic agents and spaces. The "Water Paradise" in the cultures of Veracruz Chantal Huckert. In: Budar L, Ladrón de Guevara S (Coord.) Use and representation of water on the Gulf Coast. Universidad Veracruzana-Instituto Literario de Veracruz, pp 321–344

Kloster K (2017) Social and environmental violence: the dismantling of communities in the region of Poza Rica, Veracruz, pp 183–203. In: Ruiz JLM, Licea DM, Paré L (Coords.) Conflictos por el agua y alternativas en los territorios indígenas de México. Mexican Institute of Water Technology, p 300

Lazos E, Paré L (2000) Indigenous glances towards a saddened nature. Perceptions on environmental deterioration among nahuas of southern Veracruz. Plaza y Valdés, p 230

Lazos Chavero, Paré Ouellet (2016) Miradas indígenas sobre una naturaleza entristecida. Percepciones del deterioro ambiental entre nahuas del sur de Veracruz. Edit. Plaza y Valdés-IISUNAM, México

López Austin A (2012) Cosmovisión y pensamiento indígena. UNAM-IIHS, p 15

Lorente Fernández D (2006) Nahua childhood and transmission of the cosmovision: the ahuaques or rain spirits in the Sierra de Texcoco (Mexico). Boletín Antropología Universidad Antioquia 20(37):152–168

Lorente Fernández D (2008) Rain deities, graniceros and therapeutic offerings in the Sierra de Texcoco. Anales Antropología 42(1):167–201

Lorente Fernández D (2009) Graniceros, the lightning ritualists in Mexico: history and ethnography. Cuicuilco 16(47):201–223

Lorente Fernández D (2011) La razzia cósmica. A Nahua conception of climate. Water deities and graniceros in the Sierra de Texcoco. CIESAS, UIA, p 230

Lorente Fernández D (2017) Those who understand time. Graniceros de la sierra de Texcoco. Investigaciones Sociales 20(36):69–79

Markens R (2013) The meaning of the Greca Escalonada in the prehispanic imaginary of Oaxaca: a basis of political power. Cuadernos Del Sur: Revista Ciencias Sociales 35:67–82

Martínez FX, Velasco AP (2014) Guía para la Defensa Comunitaria del Agua en Puebla y Veracruz. Fondo para la Comunicación y la Educación Ambiental. A.C.Centro Mexicano de Derecho Ambiental, A.C. Mexico, p 62 pp.

Martínez JL (2017) Presentation. In: Ruiz JLM, Licea DM, Paré L (Coordinators) Conflictos por el Agua y Alternativas en los Territorios indígenas de México. Mexican Institute of Water Technology, Mexico, p 317

Martínez Villaroya J (2019) Cosmovisión o cosmoescucha? La preeminencia del ojo en los estudios sobre Mesoamérica. Acta Poética 40(2):77–97. https://doi.org/10.19130/iifl.ap.2019.2.857

Moreno Arriba J (2016) Natural resource management in the sierra de Santa Marta in Veracruz, Mexico. An example of intercultural de(s)colonial collaboration for the sustainability of indigenous peoples in Mesoamerica. Revista OPSIS (Semiannual publication of the Unidade Acadêmica Especial História e Ciências Sociais da Universidade Federal de Goiás/Regional Catalão, Brazil) 16(1):106–130

Moreno Arriba J (2019) Alternatives to development versus bad development in the management of natural resources in the Los Tuxtlas Biosphere Reserve (Veracruz, Mexico), Fyeguyia. J Soc Territorial Stud 2(2):48–67

Murillo Licea D (2017) Una propuesta a partir de observar lo local como explosión de alternativas de manejo del agua. In: Luis J, Ruiz M, Licea DM, Paré, L (Coord) Conflictos por el agua y alternativas en los territorios de México. IMTA, pp 247–273

Ochoa Ávila MG, Arias F (2016) When Maamlaab and Junkil aab wake up: water, identity and oral tradition among the Teenek of the Huasteca potosina and Veracruz. In: Agua en la cosmovisión de los pueblos indígenas en México. CONAGUA

Oropeza Escobar M (1998) Juan Atzkin y el diluvio. Una aproximación estructural al mito totonaco. Editora Instituto Nacional Indigenista

Pacheco Mamone A, Robles Guadarrama (2014) Complejidad, resiliencia y memoria. Aportes para la comprensión de un proceso integral de restauración ecológica en la sierra de Santa Marta, Veracruz. En: COMBIOSERVE, Community Conservation in Latin America: Innovations in Research and Practice took place in the Mexican town of Xico, in the state of Veracruz, 6–9 November 2014. Global Diversity Foundation. https://old.global-diversity.org/resources/programme-reports/proceedings-combioserve-conference-memorias-congreso-combioserve/

Pacheco Mamone A, Rosario Bautista JA, Robles Guadarrama C, Cárcamo García J (2018) Reflections on a community experience. Network of weather stations in the Santa Martha mountain range. Data and perceptions. In: Memorias XXVII National congress and XII International congress of meteorology. Xalapa

Pacheco Mamone A (2013) An approach to worldviews, in popolucas communities of the Sierra de Santa Martha, Veracruz. Ph.D. Thesis, COLMICH

Pacheco Mamone A, Robles Guadarrama C (2023) Miradas desde la complejidad, comunidades indígenas, impactos y resiliencia ante el cambio climático. Del Oficio, Revista Con-temporánea. Toda La historia en el presente. 1ª primera época. vol 10, N°19–20, pp 109–127. Vista de Miradas desde la complejidad. Comunidades indígenas, impactos y resiliencia ante el cambio climático (inah.gob.mx)

Paré L, Robles C (2006) Los retos del ordenamiento comunitario en un contexto de apropiación individual del territorio. A case study in southern Veracruz. In: Fonseca A, Arreola Muñoz S, González Ortiz MA, Acosta González J (Coord) Ordenamiento Territorial Comunitario, pp 171–191

Paré L, Robles C (2008) Territorial water management in the Texizapan-Huanztlán subbasin, southern Veracruz. In: Watershed management and environmental services. Community and citizen perspectives. Ed SEMARNAT-INE-ITACA-RAISES-SENDA, A.C.-WORLD WIDLIFE FUND-GAIA, pp 139–163

Paré Ouellet L, Gerez P (2011) Al Filo del Agua: Cogestión de la subcuenca del río Pixquiac. Ed. National Autonomous University of Mexico (UNAM) Sendas, A.C. Universidad Veracruzana Secretaría de Medio Ambiente y Recursos Naturales (Semarnat) Instituto Nacional de Ecología (INE) Universidad Iberoamericana, campus Puebla Juan Pablos Editores, p 355

Paré L (2009) Social conflicts around water in Veracruz. La palabra y el hombre n 37, pp 32–37

Paré L (2017) Watershed co-management: organization, territorial integration and conflict prevention. In: Luis J, Ruiz M, Licea DM, Paré L (Coord) Conflictos por el agua y alternativas en los territorios de México. IMTA

Peña de la Paz FJ, Herrera EP, Granados LEM (2010) Indigenous peoples, local water and conflicts. In: Cisneros BJ, Torregrosa ML, Aguilar ALA (eds) El agua en México: cauces y encauces. Mexican Academy of Sciences, pp 455–478. ISBN 978-607-95166-1-1

Pérez Apango Y (2020) The ritual landscape in the dreams of southern huasteca specialists. Mirada Antropológica 15(1):75–91

Robles Guadarrama C, Pacheco Mamone A, Velasco Vázquez ML, Pereda Rivera E, Olmos Zepeda JR, Ángel Santiago A (2016) Preliminary results of the riparian diagnosis in seven ejidos of the Rio Huaunztlán watershed. In: Memorias del IV Congreso Nacional de Cuencas hidrográficas, INECOL-Universidad Veracruzana

Robles Guadarrama C, Pacheco Mamone A (2013) Sustainability and the faces of modernity in the Sierra de Santa Marta. In: Conde Flores A, Ortiz Báez P, Delgado Rodríguez A, Gómez Rábago F, (eds) Nature-society reflections from complexity. Universidad Autónoma de Tlaxcala. Centro de Investigaciones Interdisciplinarias sobre Desarrollo Regional, pp 45–483

Robles Guadarrama C (2008) Territory, water and society in an indigenous region of southern Veracruz. Master's Thesis in Rural Development, Universidad Autónoma Metropolitana, Xochimilco, Mexico

Roca Guzmán ME (2019) The origin and development of public water policy in Acayucan, Veracruz. In: Alberto D, Platas F, Sánchez IO, Buss G (eds) Agua. Territorialities and dimensions of analysis. Resistencia SA de C, pp 303–339

Romero Vivas I (2003) El agua en la cosmovisión totonaca. Boletín del Archivo Histórico del Agua, pp 48–54

SEMARNAT (2016) Spatial analysis of the most vulnerable regions to droughts. CONAGUA-SEMARNAT, p 43

Trejo Barrientos, L, González González M, Heiras Rodríguez, Lazcarro Salgado I, Sosa Fuentes S (2013) Especialistas del umbral. Don e intercambio en el chamanismo del Sur de la Huasteca. En: Los sueños y los días Chamanismo y nahualismo en el México actual. Coord. Miguel A.Bartolomé, Alicia B. Barabás. Editorial Instituto Nacional de Antropología e Historia INAH

Vargas Montero G (2010) La cosmovisión de los pueblos indígenas. Edit SEV

Venter ML (2020) Representations of water in Totogal ceramics during the Late Postclassic. In: Budar L, Ladrón de Guevara S (Coord.) Uso y Representación del agua en la costa del Golfo. Universidad Veracruzana-Instituto Literario de Veracruz, pp 363–377

Vidriales Chan G (2016) Analysis of water management in the city of Xalapa: contributions for the construction of new models of water resource management. Master's Thesis, Ed. COLVER, p 151

Villagómez Y, Gómez E (2020) Water resources in the indigenous regions of Mexico. Region Soc 1–26. ISSN e-2448-4849. https://doi.org/10.22198/rys2020/32/1288

Villagómez Y, Amoroz I, Gómez E (2013) Water resources in the indigenous regions of Mexico. El Colegio de Michoacán, AC, p 121

Chapter 8
Historical Trajectory of Water Uses in the Upper Paraná River Basin (Brazil)

Carolina Buso Dornfeld, Maria Angela de Moraes Cordeiro, Maurício Augusto Leite, and Bruna Santos Cardozo

Abstract Brazil is located in South America, has 12% of the fresh water available on the planet, and is in first place of countries with the highest amount of this resource. The Paraná River is the second largest river in Brazil, 4,880 km long, with a drainage basin of 820,000 km^2. The Upper Paraná River Basin is part of La Plata River Basin and occupies 10% of the Brazilian territory, being characterized as the most populous region with a great demand for water. Studying the distribution of water resources related to the occupation of the territory is important for its sustainable use. Based on a review of the literature, we observed that this river was initially used as a route to transport silver from Bolivia, but several native peoples lived on its banks and used its waters for navigation, fishing and irrigation. Currently, its waters are used for human and animal consumption, navigation, electricity generation, leisure, and fishing. The study of land and water use from the original peoples to the present day shows how this region was developed and will contribute to understanding the social, economic, political, and environmental relations of different historical periods and the problems associated with water in contemporary times.

Keywords La Plata River Basin · South America · Water resources · Native people

8.1 The Hydrographic Basin of the Upper Paraná River

Brazil is a country located in South America and has 12% of the freshwater available on the planet, and it is the first place among countries with the highest amount of this resource. The Upper Paraná River Basin (UPRB) is part of the La Plata River Basin (South America), which has 3,100,000 km^2 that includes Brazil, Argentina, Bolivia, Paraguay and Uruguay (Fig. 8.1). The UPRB encompasses six units (states) of the Brazilian Federation: Goiás (GO), Mato Grosso do Sul (MS), São Paulo (SP),

C. B. Dornfeld (✉) · M. A. de Moraes Cordeiro · M. A. Leite · B. S. Cardozo
São Paulo State University, School of Engineering, Ilha Solteira, Brazil
e-mail: carolina.dornfeld@unesp.br

G. A. Corzo Perez et al. (eds.), *Aqua Vitae*, The Latin American Studies Book Series,
https://doi.org/10.1007/978-3-031-57731-4_8

Minas Gerais (MG), Paraná (PR), Santa Catarina (SC), and Brasília/GO, the country's capital. The UPRB has great importance in the national context and is located in a very populous region with great economic development in the country.

Aiming at the guidance, planning and management of water resources throughout Brazil, the National Water Resources Council, which is a collegiate, consultative, and deliberative council, created by Law No. 9433/1997, National Water Resources Policy (Brasil 1997), known as the Water Law, established the National Hydrographic Division. Basins are groups of basins or nearby hydrographic subbasins, with similar social, economic, and natural characteristics are considered a hydrographic region.

The UPRB region occupies 10% of the Brazilian territory, and the Paraná River is the second largest river in Brazil, with 4,880 km in length, being formed by the Grande and Paranaíba rivers, with a hydrographic basin of 820,000 km^2 and an average flow of 1,888 m^3.s^{-1} (Etchebehere et al. 2005).

In Fig. 8.2, we can observe the area covered by the UPRB that will be the object of this investigation. The UPRB is located on the Brazilian border, and it is composed of 14 main rivers, being named:

Fig. 8.1 Boundaries of the La Plata River Basin and the Paraná River in South America (Coordinates not provided) (*Source* TFDD – Transboundary Freshwater Dispute Database 2004)

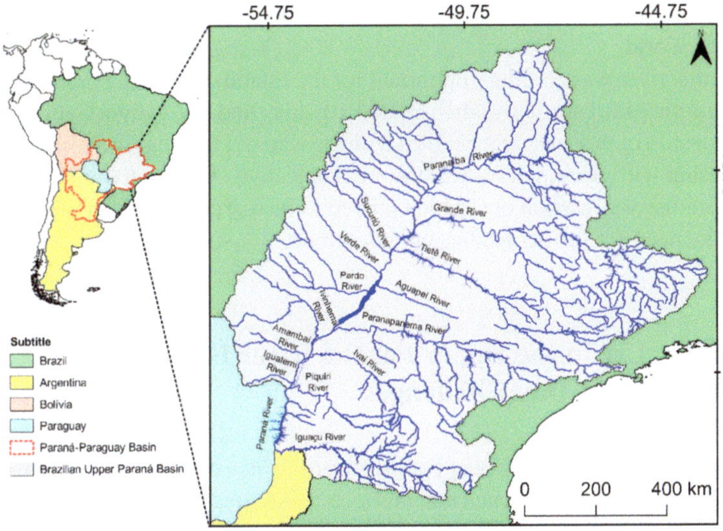

Fig. 8.2 South America, demonstrating the coverage area of the Paraná-Paraguay Basin, with emphasis on the Upper Paraná River Basin (Coordinates not provided). (*Source* Celestino et al. 2019)

- Paranaíba and Grande Rivers, which forms Paraná River;
- Right bank: Sucuri River, Verde River, Pardo River, Ivinhema River, Amambaí River, and Iguatemi River.
- Left bank: Tietê River, Aguapeí River, Paranapanema River, Ivaí River, Piquiri River, and Iguaçu River.

According to Rudke (2018, p. 32), UPRB comprises approximately one-third of the Paraná River Basin, 898,880 km^2, of which 97% belong to "Brazilian territory," and this region has great importance for the country, as it has high occupancy rates (35% of the population). The UPBR has large areas of industries, agriculture, livestock, and historic dam constructions; consequently, few areas are still found in original conditions (Agostinho et al. 2007).

In 2022, 64,322,182 people lived in the region, with 93.0% in urban areas (IBGE 2022). The region has the most populous city in South America, São Paulo, with 11.5 million people. Other important populations centers are: Brasília, Curitiba, Goiânia, Ribeirão Preto, Campo Grande, and Uberlândia. Most of the population is concentrated in the Units Hydrographics of the Tietê and Grande rivers, which together correspond to 61.0% of the total population and the highest demand for water resources in the country.

In addition, the Paraná River integrates the scenario of the trinational border between Brazil, Paraguay, and Argentina. Its navigability presented the possibilities to establish commercial and communication relations between countries. According to data from Itaipu Binacional (2022), it has a total extension of 4,880 km to the mouth of La Plata River, in the city of Buenos Aires, considered the eighth largest river in the world.

Thus, this river was certainly important for the establishment and organization of precolonial peoples, especially when related to fishing and navigation. Understanding the relationship of these ancient peoples with the water of this region and establishing a relationship with current times was the main objective of this text. Finally, a review of the literature was carried out, which proved to be scarce and which will certainly awaken the need for further studies on this subject.

8.2 Water Use in the Upper Paraná River Basin: Past and Present-Day Challenges

To understand the role of the Upper Paraná River Basin in the settlement of humans in South America, it can be assumed that this area constituted one of the main migratory routes of prehistoric hunter-gatherer-fishermen. The foods consumed were obtained according to natural and seasonal availability. Furthermore, these people lived in camps characterized by short stays and permanent rotation in large areas, a typical phenomenon of nomadism (Kashimoto and Martins 2005).

On the border of Spanish and Portuguese territories from the sixteenth century onwards, the Upper Paraná was the scene of navigation and strategic occupations. Currently, with the construction of several reservoirs and dams for the production of hydroelectric energy, most of these landscapes have been changed in environmental and in occupation terms.

Originally, according to Portz et al. (2020, p. 38), the entire area of "Brazilian extension, through Paraguayan territory until reaching the province of Misiones in Argentina, was covered by dense and humid subtropical forests, where araucaria, cedar, ipê, peroba and, in a large part of the region, the native yerbamate." Alberto Daniel Alcaráz, in his book "La investigación y las actividades económicas en el alto Paraná (1880–1920)," brings discussions about economic activities and their relationship with transport in Argentine territory. The author reports how the soil fertility around Foz do Iguaçu and the natural wealth boosted agricultural crops. In addition, it brings several observations regarding transport, with proposals for increasing waterways around the area, mainly of the weed, matt, and wood (Alcaráz 2010).

Research carried out with excavations of 24 archeological sites, especially on the right bank of the Paraná River, identified traces of different populations of ancestral peoples, with a total of 50,000 artifacts being collected in these locations (Kashimoto and Martins 2005), concluding that prehistoric peoples, with traces dating from 4000 to 6000 years ago, were formed by human groups of hunter-gatherer-fishermen. These records indicate the human presence in the Pleistocene, comprised in the Quaternary period, the Cenozoic era. According to the researchers, among the lithic artifacts found, hundreds of scrapers, planes, knives, hand axes, choppers, and chopping tools were recorded, suggesting the use of these tools for the construction of canoes and ornaments, as well as the processing of fish, as abundant in this river course (Kashimoto and Martins 2005, p. 57).

In relation to indigenous peoples, traces of about 1000 years ago were found, featuring agricultural, semi-nomadic (Amaral 2019, p. 22), and ceramic producers, mainly from the *Kayapó* and *Tupi Guarani* ethnicities. These traces were found only 40 cm deep. At that time, the *Tupi Guarani* became hegemonic in the area, since environmental conditions allowed their permanence, such as the increase of organic matter in soils, which provided the development of richer soils and the expansion of the semi-deciduous alluvial forest and its associated fauna. It was like this, the occupation of the right banks of the Upper Paraná on the eve of the Portuguese colonization process in Brazil in 1500.

According to Portz et al. (2020, p. 40) from travelers reports who passed through the trinational border, moving through the waters of the Paraná River, considered the most important artery of communication and transport in the region during the colonial period. So, at first, the waters of the Paraná River also served as a route for the transport of silver from Hispanic America, and several native people lived along its banks. of river According to the same author, chronicler records portray that the surroundings of the Paraná River were mostly occupied by Guarani peoples, indigenous populations such as the *Tupi*, *Kaingang*, *Gualachos*, and *Guayra*, on the Brazilian side, and the *Guaná* and *Gauyaki*, on the Paraguayan side of the basin, appear in the records of historian and anthropologists (Portz et al. 2020, p. 39), and these peoples lived and used its waters for navigation, fishing and irrigation.

The natives continued to occupy the banks of the Paraná River, although most of the indigenous population disappeared after European colonization, and with the subsequent occupation of lands to the west of Brazilian territory. The *Tupi Guarani* still resist and occupy some localities near the Paraná River, especially in the state of Paraná (Amaral 2019, p. 26). However, as a product of a history of fulminant interethnic conflicts, unfortunately, at present, there are no longer any indigenous Guarani living on the banks of the Upper Paraná (Kashimoto and Martins 2005, p. 62).

Despite the entire process of inculturation of Christianity, carried out by Portuguese Jesuit priests, several terms and names related to water resources, which had been named by indigenous peoples, were maintained (ANA 2007). Some examples are igarapés (narrow natural channel navigable by small boats, which is formed between two river islands or between a river island and mainland), paranás (branch of a mighty river separated from it by a series of islands), and ipueiras (pond formed by the overflow of rivers in low places) (DICIO 2022).

However, the advance to the west, from the Atlantic Ocean to the Paraná River, after the colonial period, is marked by intense conflicts between the indigenous population, explorers in search of precious metals (called the *Bandeirantes Paulistas* Movement) and European and Asian immigrants, who were looking for work and land.

In addition, we have the record of the voyage of Teotônio José Juzarte, in 1769, who traveled and mapped the Tietê, Paraná, and Iguatemi rivers. After two hundred years, journalist Theophilo de Andrade, in his first contact with the Paraná River, observed that despite the abundance of water, forest, fish, and terrestrial animals, the region was almost uninhabited, as they suffered from mosquitoes and malaria, in addition to intense heat, conditions that apparently only the natives could withstand (Kashimoto and Martins 2005).

Anyway, due to its navigation potential, Paraná River was used in several Brazilian historical moments, as mentioned by Kashimoto and Martins (2005), such as:

(1) Way of escape of the Jesuit priests and the *Tupi Guaranis* toward the southern region of the country, when they were pressured by the advance of the Bandeirantes Paulistas Movement. This movement, according to Amaral (2019, p. 28), which took place at the beginning of the seventeenth century, was looking for precious metals and cheap labor in the western state of São Paulo. Many atrocities were committed, which led the Spanish Jesuits to write a letter denouncing the horrors to Pope Urbano VIII, who published a pronouncement excommunicating all those who sought to imprison indigenous people. However, the movement continued for many years.
(2) Route used by the *Bandeirantes Paulistas* Movement in their incursions into the west of the country.
(3) Interconnection between the Tietê and Pardo rivers, in the movement called the mining cycle (which lasted until the end of the nineteenth century), and also toward the livestock regions in the state of Mato Grosso (beginning of the twentieth century).

During this period, according to Amaral (2019, p. 29), human occupation on the banks of the Paraná River was associated with the creation of communication routes (bottle roads and river ports), which would transport cattle from Mato Grosso do Sul to the coast of the states of São Paulo and Rio de Janeiro. In this period, there was also the expansion of coffee culture in this region, which stimulated the expansion of the railways. This expansion attracted populations to the west of São Paulo, causing several conflicts with indigenous populations.

Also, at the beginning of the twentieth century, European and Asian immigrants (Hungarians, Japanese, Czechs, and Germans), attracted by the construction of ports and the railway, founded several riverside villages. Life on the banks of the Paraná River in the first half of the twentieth century was an intense relationship with the river in its natural state: waterfalls, currents, islands, and sinuous curves (Amaral 2019, p. 31).

From the 1950s onwards, there was intense port activity in the Paraná River for the flow of hardwoods (Kashimoto and Martins 2005, p. 43), and from 1970 onwards, the Paraná River became the subject of studies for its hydroelectric use. So, in addition to being used for human and animal consumption, navigation, energy generation, leisure, and fishing, it still carries an important historical and cultural aspect.

8.3 Water Uses in the Upper Paraná River Basin in the Twentieth and Twenty-First Centuries

In Brazil, the main water uses are irrigation, supply for urban and rural human use demands, industrial, livestock, sand mining, power generation by thermoelectric and hydroelectric plants, waterway, and aquaculture (ANA 2019).

The UPRB has the highest consumption of water (496 $m^3.s^{-1}$) in the Paraná Basin, with 38.3% destined for irrigation, 35.8% urban human, 1% rural human, 19.3% industrial, 8.2% livestock, 1.7% thermoelectric, and 0.2% mining.

In a historical point of view, all uses are connected with the development of Brazil, mainly at the end of the nineteenth century.

Power generation is one of the main needs of the modern world, and hydroelectric energy appears as an alternative to meet this demand, according to data from the National Electric Energy Plan (DNAEE 1995). Magalhães and Tomiyoshi (2011) report that a Geographical and Geological Commission of the State of São Paulo led expeditions to assess natural resources such as waterfalls and rivers in 1890 and 1910, seeking hydroelectric potential. These expeditions really confirmed the potential for electrical energy to be produced. The first hydroelectric plant in the State of São Paulo was installed on the Monjolinho River in 1893, and between 1890 and 1930, 118 hydroelectric plants were installed on its territory (Nogueira and Argollo Ferro 2015).

According to the National Electric Energy Agency (ANEEL 2003), hydroelectric projects greater than 30 MW are considered as Large Power Plants (LPP), greater than 1 MW and equal to or less than 30 MW are Small Hydropower Plants (SHP), and plants with an installed capacity of up to 1 MW are called Hydroelectric Generating Centers (HGC). The production of hydroelectric energy according to Müller (1995, p. 17) consists of harnessing the force of the waters of a river or reservoir to turn the hydraulic turbines, which drive the generators, producing electric energy.

The falls on the Upper Paraná River, that, from their spectacular character, have received most attention, are those at Guayra and Iguazu (Barclay 1932). The UPRB has seven of the ten largest hydroelectric plants in Brazil: Itaipu (Paraná River—7,000 MW), Ilha Solteira (Paraná River—3,444 MW), Itumbiara (Paranaíba River—2,082 MW), São Simão (Paranaíba River—1,710 MW), Governador Bento Munhoz da Rocha Neto (Iguaçu River—1,676 MW), Jupiá (Paraná River—1,551.2 MW) and Porto Primavera (Paraná River—1,540 MW). The UPRB has an installed capacity of 43,635.56 MW according to the Brazilian Hydroelectric Potential Information System (SIPOT, 2018), which represents 40.62% of Brazilian Hydroelectric Energy.

The hydroelectric power plants used in the Upper Paraná River Basin were shown in Fig. 8.3 and are possible to observe that during the 1940s, the predominant use for water in all Brazilian states was for human consumption (rural and urban). The industrialization process prompted a parallel increase in industrial use through the mid-1980s, when stagnation held sway, followed by an upswing in expansion during the first decade of the second millennium. A slight decline in growth has been observed since 2012, related to the water crises that hit several regions in Brazil, in addition

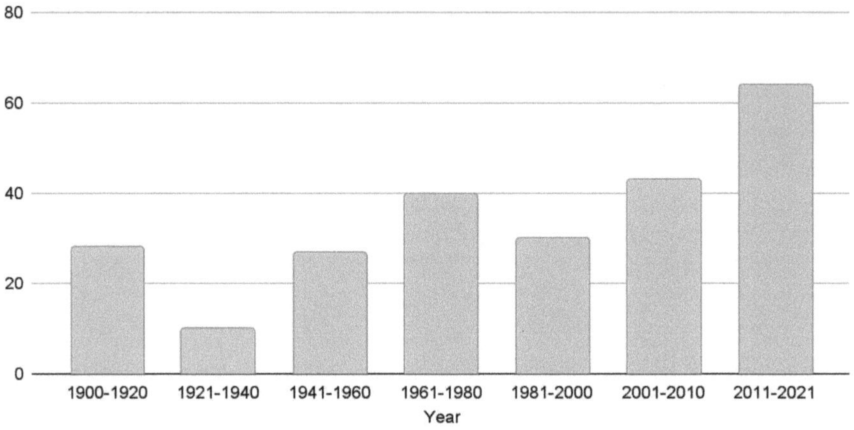

Fig. 8.3 Hydroelectric power plants in use in Upper Paraná River Basin (*Source* Created by the authors using data from ANEEL, 2022) Data: Enterprises database—Generation Information System; ANEEL—National Agency of Electric Energy—Sum of LPP, SHP and HGC in Paraná River Basin

to the slowdown in Brazil's economic progress. From this year, an increase was also noted in the water used to generate thermoelectric energy, prompted by the need to bring complementary energy sources into operation due to low water volumes available for hydropower generation in Brazil.

The UPRB has 1,507 cities, 256 have more than 50,000 inhabitants (IBGE 2022), including: São Paulo/SP (11,451,999), Brasília/DF (2,817,381), Curitiba/PR (1,773,718), Goiânia/GO (1,437,366), Campo Grande/MS (898,100), Dourados/MS (243,367), Ribeirão Preto/SP (698,642), Uberlândia/MG (713,224), Uberaba/MG (337,836) and Londrina/PR (555,965). The water use data for these cities were compiled from a database of the National Water and Sanitation Agency (ANA 2019) to prepare Fig. 8.4, demonstrating that from the end of the nineteenth century to the present day, the UPRB presented a population growth that increased water consumption.

The data showed the exodus of the population from the rural area to the urban area and the increase in urban water consumption and industry. As a result, there was an increase in population density (73.4 inhabitants.km^{-2}) and the largest gross domestic product in Brazil (Lima et al. 2016).

Other water uses, such as sand mining and thermoelectric plants, had less expressive uses of water. In 1931, mining and thermoelectric plants consumed, respectively, 3.5×10^{-4} and 0.0 $m^3.s^{-1}$, and in 2021 they consumed 0.0315 $m^3.s^{-1}$ and 0.025 $m^3.s^{-1}$. Regarding irrigation use, in Brazil, irrigation began between the end of the nineteenth century and the beginning of the twentieth century in rice plantations in Rio Grande do Sul. In the 1960s, new irrigation poles were consolidated in the states of São Paulo and Minas Gerais. The most irrigated crops are rice, coffee, and fruits, and the typology that leads the growth of irrigation is pivot systems (Fig. 8.5), with an efficiency of 85% (ANA 2021). According to ANA (2013), the irrigated area of the

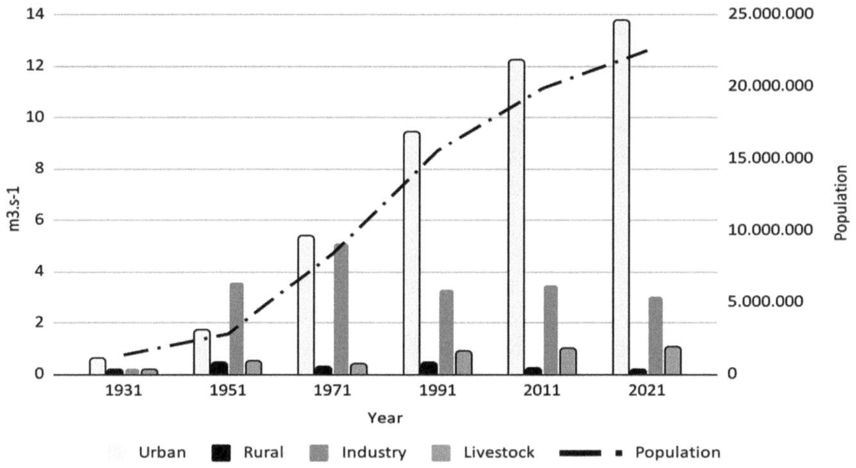

Fig. 8.4 Water consumption in ten cities in the Upper Paraná River Basin (*Source* Created by the authors using data from ANA, 2019). Cities: São Paulo/SP, Brasília/DF, Curitiba/PR, Goiânia/GO, Campo Grande/MS, Dourados/MS, Ribeirão Preto/SP, Uberlândia/MG, Uberaba/MG e Londrina/PR

UPRB is 2,106,232 hectare, which corresponds to 36.3% of the 5,800 km^2 irrigated hectare in Brazil, and irrigation is mainly concentrated in the states of São Paulo, Goiás, and Minas Gerais.

Irrigation uses 189.96 $m^3.s^{-1}$ of water in the UPRB; however Guimarães and Brandão (2020) mentioned that agriculture was responsible for 25% of Brazil's gross

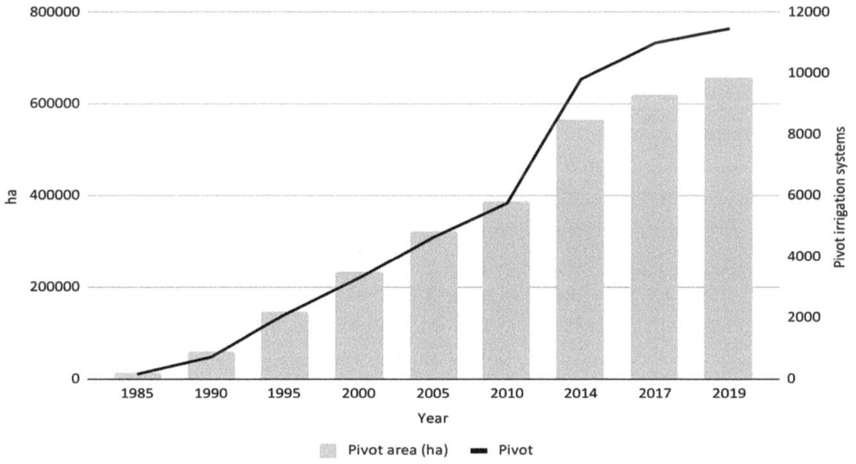

Fig. 8.5 Pivot irrigation systems in Upper Paraná River Basin (*Source* Prepared by the authors using data from ANA, 2022)

domestic product, with 35% of the workforce, and contributes 40% to the Brazilian trade balance.

The National Water and Sanitation Agency (ANA 2021) emphasize that recreational water use needs legal requirements and management instruments, such as granting the right to use of water resources (authorization for the use of water) and the charge for the use, to ensure the sustainability of the activity, increasing efficiency and reducing the waste of water.

Use for navigation on the Tietê and Paraná rivers began even before Portuguese colonization in Brazil, as the indigenous people already made river crossings and used the fish for food—it has been the possibility of plural water use since then. At the end of the nineteenth century, the Tietê River began to be explored by expeditions (1774 and 1778), and in the nineteenth century, scientific commissions traveled the river seeking to verify the navigability conditions (Mello et al. 1990).

Transport along the Tietê and Paraná rivers, until the mid-twentieth century, was difficult due to natural conditions and relief topography. Only with the technical improvement and engineering, the rivers were useful for navigation. So, in the 1940s and 1950s, the idea of creating the Paraná-Tietê waterway system emerged, characterized by territorial integration through the port and waterway infrastructure, creating the necessary conditions for long-distance transport.

Thus, through various engineering works, based on the principle of multiple use of river courses (electricity production and transport) and given the demographic and economic demand of the center-south of the country, the material bases were created for the emergence of the Tietê-Paraná Waterway. The Tietê Waterway began in 1981, with the regional transport of sugarcane. In 1991, the long-distance service began, comprising the Tietê section and the north section of the Paraná River. The union of the two river courses constitutes the Tietê-Paraná waterway, which is 2400 kilometers long (1600 km referring to the Paraná River and 800 km the Piracicaba and Tietê rivers) (Felipe Junior 2008). The main products transported are sulfur, rocks, gypsum and lime; soy; corn; sugar; and fertilizers.

The UPRB supplies the most populous and industrialized regions of Brazil and has multiple use reservoirs, used for fish farming in net cages. Currently, there is regularized aquaculture in 18 reservoirs called: Água Vermelha, Canoas I, Canoas II, Chavantes, Capivara, Igarapava, Ilha Solteira, Itaipu, Itumbiara, Jaguará, Jupiá, Jurumirim, Porto Primavera, Rosana, Salto Caxias, Segredo, Taquaraçu, and Volta Grande.

Fish farming in net cages is a very recent activity, which began in the 1980s and has great growth potential, especially in reservoirs. Fish production in Brazil in 2019 was 529,623 thousand tons, with Tilápia (*Oreochromis spp.*) being the species responsible for 61% of this production (Fig. 8.6).

The increase in the amount of water and the growth of water uses in the Upper Paraná River Basin over the years demonstrated the importance for the development of Brazil. However, the growth of the population and activities in this basin has brought some challenges that will have to be faced in the next few years.

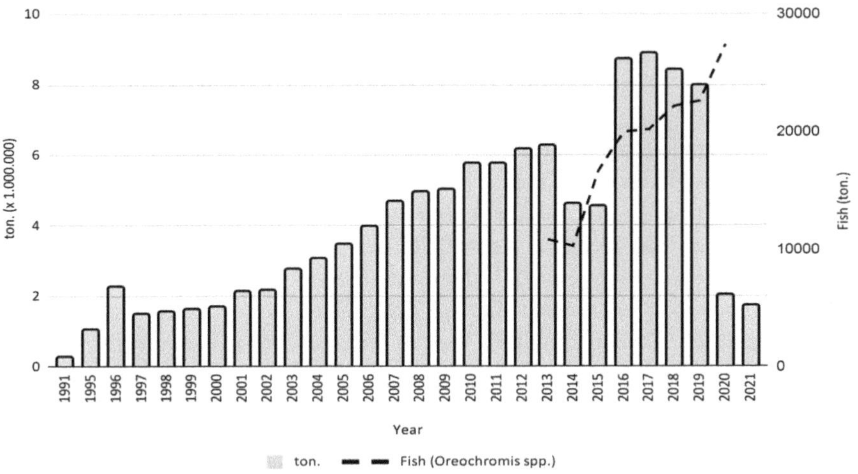

Fig. 8.6 Waterway transport and aquaculture in eight reservoirs from the Upper Paraná River Basin (*Source* ANA 2019; IBGE 2020; Bigaran and Tizato 2009). Reservoirs: Ilha Solteira, Jupiá, Jaguara, Igarapava, Água Vermelha, Jurumirim, Chavantes e Canoas I)

8.4 Challenges for the Upper Paraná River Basin

Some historical aspects of the Paraná River and the UPRB showed fundamental importance in the population of South America. In addition, the emergence of cities, water supply, electricity generation, food and agriculture, livestock, irrigation, and transport boosted the development and growth of Brazil.

All of these water uses make the territory of this basin extremely dependent on the quantity and quality, directing the stakeholders to the challenge of water resources management.

The rise and fall of civilizations often appear related to the use and abuse of water, and societies were called hydraulic civilizations due to the development of agriculture and society linked to the control of water, especially when it comes to irrigation.

Veldkamp et al. (2017) mention that movements into water scarcity due to human impact are predominantly found in more downstream regions, whereas people moving out tend to live in relatively upstream areas apart from the Paraná where those moving into water scarcity live relatively more upstream. While upstream water consumption is the dominant driver of changes in water availability, land use and land cover change and reservoir operations dominate the change in water availability in the Paraná Basin year-round. Almost always, deforestation plays a role in water deficits. Rocha (2010, p. 193) states that the loss of tree cover, in the short term, reduces the loss of soil water through transpiration, as well as causes greater water flow on the soil surface, since the old dampening blanket of fallen leaves has been replaced by bare earth. It is now also known that land use for agricultural production, urban development, the use of groundwater for irrigation or supply, and the construction of dams for energy production also contribute to changes in the hydrological cycle.

Due to annual periods of drought and rain and changes in hydrological cycles, evidence is accumulating that the climate is changing in different parts of the planet.

In the twenty-first century, the UPRB presented a water crisis in the years 2000, 2014 and 2021 that caused a decrease in the Paraná River water level. The results presented in the Final Technical Report "Reduction of Outflow" (São Paulo, 2021, p. 194), with focus on the year 2021, demonstrated low values in water precipitation in recent years and a very severe shortage represented by low water levels in reservoirs and restrictions on the multiple uses of these resources, including navigation, irrigation, leisure and tourism, fishing, urban supply and power generation. In view of the risks of aggravation of the problems, a committee for the management of the water crisis was created at the federal government level and proposed, among other measures, greater control of effluent flow, aiming to ensure navigation and future energy supply.

Thus, there is a new scenario: on the one hand, there is an increase in population, agricultural and industrial activities, and on the other hand, variations on precipitation that harm activities in this basin. The alternative will be for the stakeholders to understand that the sustainable use of water will be a solution to the maintenance of the population and environmental quality.

References

Agostinho AA, Pelicice FM, Petry AC, Gomes LC, Júlio HF (2007) Fish diversity in the upper Paraná River basin: habitats, fisheries, management, and conservation. Aquat Ecosyst Health Manage 10(2):174–186. https://doi.org/10.1080/14634980701341719

Alcaráz AD (2010) La investigación y las actividades económicas en el alto Paraná (1880–1920). Misiones, Argentina, p 68. Retrieved 10 February 2022, from https://ri.conicet.gov.ar/handle/11336/123880

Amaral AC (2019) Pelos caminhos do rio Paraná. Editora Horizonte, São Paulo, p 144

ANA. Agência Nacional de Águas (Brasil) (2007) History of the use of water in Brazil: from the discovery to the 20th century. ANA, Brasília, p 249

ANA. Agência Nacional de Águas (Brasil) (2013) Conjuntura dos recursos hídricos no Brasil: 2013. ANA, Brasília, p 432

ANA. Agência Nacional de Águas e Saneamento Básico (Brasil) (2019) 18 - Manual de Usos Consuntivos da Água no Brasil — Sistema Nacional de Informações sobre Recursos Hídricos. (n.d.). Retrieved 13 February 2022 from http://www.snirh.gov.br/portal/snirh/centrais-de-conteudos/central-de-publicacoes/ana_manual_de_usos_consuntivos_da_agua_no_brasil.pdf/view

ANA. Agência Nacional de Águas e Saneamento Básico (Brasil) (2021). Atlas irrigação: uso da água na agricultura irrigada. ANA, Brasília, p. 130. Retrieved 15 February 2022, from https://portal1.snirh.gov.br/ana/apps/storymaps/stories/a874e62f27544c6a986da1702a911c6b

ANA. Agência Nacional de Águas e Saneamento Básico (Brasil) (2022). Agricultura Irrigada por Pivôs Centrais no Brasil (n.d.). Catálogo de Metadados da ANA. Retrieved 13 February 2022, from https://metadados.snirh.gov.br/geonetwork/srv/api/records/e2d38e3f-5e62-41ad-87ab-990490841073#:~:text=Em%202017%2C%20foram%20identificados%201%2C476

ANEEL. Agência Nacional de Energia Elétrica (Brasil) (2003). Guia do empreendedor de pequenas centrais hidrelétricas. Brasília: ANEEL. 704 págs

ANEEL. Agência Nacional de Energia Elétrica (2022). Informações de Geração da ANEEL. Retrieved February 13, 2022, from https://www.aneel.gov.br/siga#:~:text=O%20Sistema%20de%20Informa%C3%A7%C3%B5es%20de

Barclay WS (1932) The Basin of the Paraná River. Geogr J 79(2):81–99. Retrieved 10 February 2022, from https://doi.org/10.2307/1785088

Bigaran JT, Tizato LHG (2009) Hidrovia Tietê-Paraná. Retrieved January 17, 2022, from https://esalqlog.esalq.usp.br/upload/kceditor/files/2000/05/Hidrovia-Tiet%e2%94%9c%c2%ac-Paran%e2%94%9c%c3%ad-BIGARAN-JT-TIZATO-L.-H.-G.pdf

Brasil (1997) Política Nacional de Recursos Hídricos, Lei n° 9.433 de 1997. Retrieved on February 08, 2022, from http://www.planalto.gov.br/ccivil_03/leis/l9433.htm

Celestino LF, Sanz-Ronda FJ, Miranda LE, Makrakis MC, Dias JHP, Makrakis S (2019) Bidirectional connectivity via fish ladders in a large Neotropical river. River Res Appl 35(3):236–246. Retrieved 9 February 2022, from https://doi.org/10.1002/rra.3404

DICIO. Dicionário Online de Português. Retrieved 22 July 2022, from https://dicio.com.br

DNAEE. Departamento Nacional de Águas e Energia Elétrica (1995) Plano Nacional de Recursos Hídricos. Brasília, p 95. App.powerbi.com. Retrieved February 13, 2022, from https://app.powerbi.com/view?r=eyJrIjoiZmNmMmU0ZTYtYmZiYS00ZTk3LWFlMTMtZjQwMzU1MGQ1Y2E4IiwidCI6ImUwYmI0MDEyLTgxMGItNDY5YS04YjRkLTY2N2ZjZDFiYWY4OCJ9

Felipe Junior NF (2008) A hidrovia Tietê-Paraná e a intermodalidade no estado de São Paulo (p. 290) [Dissertação A hidrovia Tietê-Paraná e a intermodalidade no estado de São Paulo]. Universidade Estadual Paulista – Unesp. Retrieved on February 10, 2022, from https://repositorio.unesp.br/bitstream/handle/11449/96693/felipejunior_nf_me_prud.pdf?sequence=1

Etchebehere ML, Fulfaro VJ, Saad AR, Perinotto JAJ (2005) The Araras Escarpment in the upper Paraná River: implications to fluvial neotectonics on the Paraná drainage net evolution. Revista Brasileira De Geomorfologia 6(1). https://doi.org/10.20502/rbg.v6i1.45

Guimarães DP; Brandão GR (2020) Variação Geográfica da Agricultura Irrigada. Capítulo 51. In: Dinâmica da Produção Agropecuária e da Paisagem Natural no Brasil nas últimas décadas. Retrieved December 10, 2021, from https://ainfo.cnptia.embrapa.br/digital/bitstream/item/214965/1/Cap51-AgriculturaIrrigada.pdf

IBGE. Instituto Brasileiro de Geografia e Estatística (2022) Censo. Retrieved 12 February 2022, from https://censo2010.ibge.gov.br/

IBGE—Instituto Brasileiro de Geografia e Estatística (2020) Cidades, Brasil, Panorama. Retrieved 10 February 2022, from https://cidades.ibge.gov.br/brasil/panorama

Itaipu Binacional (2022). Rio Paraná. Retrieved 10 February 2022, from https://www.itaipu.gov.br/energia/rio-parana

Kashimoto EM, Martins GR (2005) Uma longa história em um grande rio: cenários arqueológicos do Alto Paraná. Editora Oeste Campo Grande, Mato Grosso do Sul, p. 100. Retrieved 15 February 2022, from https://muarq.ufms.br/files/2021/03/Uma-Longa-Historia-de-um-Longo-Rio.pdf

de Lima EA, Cano H, Nascimento JAS (2016) Uma Contribuição à geografia dos recursos hídricos. In: IBGE – Instituto Brasileiro de Geografia e Estatística (ed.). Brazil: uma visão geográfica e ambiental no início do século XXI. IBGE, Coordenação de Geografia, Rio de Janeiro, p. 435. Retrieved 12 February 2022, from https://biblioteca.ibge.gov.br/index.php/biblioteca-catalogo?view=detalhes&id=297884

Magalhães G, Tomiyoshi L (2011) Electricity in Brazil—Part 1. IEEE Industry Applications Magazine 8-12. Retrieved 22 July 2024, from https://doi.org/10.1109/MIAS.2010.939808

Mello MRC, Marsiglia C, Bacellar NRR (1990) História da Energia Elétrica em São Paulo - S.A. Empresa Elétrica do Itapura – CESP Companhia Energética de São Paulo, p. 88

Müller AC (1995) Hidrelétricas, meio ambiente, e desenvolvimento. Markron Books, São Paulo, p .412

Nogueira DMA, Argollo Ferro AM (2015) Hidrelétricas no desenvolvimento urbano e territorial de São Paulo. Labour & Engenho, Campinas [SP], Brasil, 9(1):19–36

Portz S, Mascarenhas M, Gregory V (2020) O Rio Paraná e a Dinâmica da Fronteira nos Séculos XIX e XX. Semantic Scholar. Retrieved 10 February 2022, from https://doi.org/10.32915/pleiade.v14i30.661

Rocha PC (2010) Indicadores de alteração hidrológica no Alto Rio Paraná: intervenções humanas e implicações na dinâmica do ambiente fluvial. Sociedade Natureza 22(1):191–211. Retrieved on February 10, 2022, from https://doi.org/10.1590/s1982-45132010000100014

Rudke AP (2018) Dinâmica da cobertura do solo para a Bacia Hidrográfica do Alto Rio Paraná, p. 110. [Dissertação - Dinâmica da cobertura do solo para a Bacia Hidrográfica do Alto Rio Paraná]. Universidade Tecnológica Federal do Paraná, Londrina. Retrieved February 10, 2022, from http://repositorio.utfpr.edu.br/jspui/handle/1/3328

São Paulo Companhia Energética de São Paulo (org.) (2021) Relatório Técnico Final Redução da Vazão Defluente UHE Eng. Sérgio Motta – Porto Primavera. Online, p 225

São Paulo (2021) Informativo DH Hidrovia Tietê-Paraná Atividades em andamento Setembro de 2021. Retrieved 10 January 2022, from http://www.dh.sp.gov.br/wp-content/uploads/2021/10/Informativo-Hidrovia-Tiete-SETEMBRO-2021.pdf

SIPOT (2018) Brazilian hydroelectric potential information system. Potencial Hidrelétrico Brasileiro. Retrieved on February 13, 2022, from https://eletrobras.com/pt/Paginas/Potencial-Hidreletrico-Brasileiro.aspx

TFDD. Transboundary Freshwater Dispute Database (2004). La Plata River Basin. Program on Water Conflict Management and Transformation. Oregon State University. Transboundarywaters.science.oregonstate.edu. Retrieved 10 February 2022, from https://transboundarywaters.science.oregonstate.edu/content/la-plata-river-basin.

Veldkamp T, Wada Y, Aerts J et al. (2017) Water scarcity hotspots travel downstream due to human interventions in the 20th and 21st century. Nat Commun 8:15697. https://doi.org/10.1038/ncomms15697

Part III
Modern Currents and Future Streams

Chapter 9
Chronology of the Use and Technologies for Water Exploitation in Cuba. Situation in a Changing Climate

Aliana López Mayea, Miriela García García, and Maria Nell González Cruz

Abstract The objective of this article is to provide information on the historical aspects, myths, beliefs, use, and management of water by Cuban society from the past to the present. Additionally, it aims to facilitate an assessment of the potential impacts of climate change on water resources, through a documentary analysis of specialized literature, consultations with publications in various media, textbooks, and written press. A documentary compilation is provided on subsistence activities and water resources at various stages of Cuban society's development. The compilation also includes considerations on the impact of climate change on these resources.

Keywords Myths · Beliefs · Climate change · Water resources · Water chronology

9.1 Background

As evidenced by articles, chronicles, and newspapers, the primary ceremonial activities of aboriginal farming societies were linked to the presence of rain, the fertility of orchards and their production, the reproductive cycles of flora and fauna, and social reproduction as a whole. These societies attributed special significance to Boinayel, the deity associated with rainfall (Arrom 1990). The truth is that the beliefs passed down by ancestors form a unique blend of culture and religion for a community. This is evident in the case of a small group of peasants known as "Los Acuáticos," who are well-known for their water rogations to bring rain.

The link between humans, water, and nature is an integral part of the cultural identity of Cuban communities. The presence or scarcity of this resource has a significant impact on management practices (Jiménez et al. 2014). In the case of Cuban aborigines, their level of development was not comparable to that of other pre-Columbian cultures, such as the Maya, Aztecs, or Incas.

A. L. Mayea (✉) · M. García García · M. N. G. Cruz
Centro Meteorológico Provincial de Ciego de Ávila, Ciego de Ávila, Cuba
e mail: anailalm@gmail.com

G. A. Corzo Perez et al. (eds.), *Aqua Vitae*, The Latin American Studies Book Series,
https://doi.org/10.1007/978-3-031-57731-4_9

In the sixth century, an indigenous group settled in the eastern region of the country. It is believed that they introduced important crops such as corn, cassava, and tobacco. As a whole, the Arawak groups of Antillean origin were known as Tainos. They were described as an agro-pottery community, but their agricultural practices were relatively primitive compared to those of other ethnic groups on the continent.

The Aboriginal settlements were located in mountainous areas, particularly near rivers or areas where multiple springs flowed, which fertilized the soil either permanently or seasonally (Fernández et al. 2012).

During the colonial period, the supply of precious water was achieved through the use of alternative methods such as capturing and storing water in tanks, including cisterns and reservoirs. Additionally, wells were constructed and moments of water scarcity were detected, all of which were essential in the founding of the colonies. One remarkable engineering feat of the mid-nineteenth century was the construction of the Albear aqueduct.

There are currently ten water diversion systems constructed, three of which stand out as the largest in the country: the Central-East Water Diversion, North–South Water Diversion, and East–West Water Diversion (Batista 2019).

Research predicts that Cuba's future climate may become more arid and extreme, with prolonged and frequent droughts and severe water deficits. The dry landscapes of the eastern zone will intensify and gradually spread throughout the rest of the country. This will result in a shift from a humid tropical climate to a dry subhumid climate, which poses a threat of desertification processes (Paz 2010).

9.2 Mythical and Religious Beliefs

None of the various sources of the Hispanic chronicles made a direct allusion to the mythical-religious beliefs of the groups settled in the Cuban archipelago, except for Las Casas (1912). Las Casas sometimes repeats what was expressed by the friar Ramón Pané, who wrote the report "Relación acerca de las antigüedades de los indios" (Arrom 1990). This report was commissioned by Admiral Christopher Columbus and refers to aspects of the mythology heard from the natives of Hispaniola themselves (Arrom 1990).

It is prudent to acknowledge the strong possibility that the social phenomenon we are currently addressing was connected to a prolonged period of environmental and climatic crisis. This period was highly unfavorable for subsistence activities such as agriculture, river, and sea fishing. It would be wise to consider the tendency of seeking solutions from a magical-religious perspective that favored the proliferation and strengthening of the cult of deities such as the rain weeper, the frog, and eventually the dolls. Ceremonialism played a significant role in this phenomenon, as suggested by the archeological evidence studied by Fernández et al. 2012.

Although Antillean aboriginal mythology features numerous deities that attempt to explain natural phenomena, their representations in iconography highlight the special significance of Boinayel, the deity associated with rainfall (Arrom 1990). Very little is known about the character Boinayel, a male deity known for his attributes as the producer of downpours that are necessary for eliminating malignant plagues and ensuring the adequate growth of plants. The indigenous plastic art depicts a rain-producing entity with tearful images, which archeologists have named "llora-lluvias" after the term coined by Don Fernando Ortiz in 1947 (Ortiz 1947a; Jiménez 1981; Fernández and González 2001a). This entity is closely associated with the act of drinking and the advancement of agriculture (Fernández et al. 2012).

In the twenty-first century, there exists a small community of peasants known as "Los Acuáticos" (the aquatic ones) who reside in the Viñales mountains of the Pinar del Rio province. This community, which has survived for several generations, believes that water has the power to cure any disease or ailment. Despite their isolation, they continue to hold steadfast to this belief. Ricardo Alvarez, a historian from Viñales, explains that the origins of this tradition can be traced back to newspaper articles from the 1930s. Interestingly, the tradition did not originate in Viñales Valley, but rather in the San Felipe Keys, a rural community also situated in Pinar del Río. According to Álvarez, Antoñica Izquierdo was able to save one of her youngest sons after a divine revelation by submerging him in a spring in the area (Sánchez 2018).

The truth is that beliefs transmitted by the ancestors make up a mixture of culture and religion of the peoples, which transited through time and arrive in these moments wrapped by the myth, forming part of everyday life, the fountain of eternal youth forming part of a myth developed over time and history in various regions of the world. The medicinal spas in Cuba offer an alternative treatment for various ailments, with locations such as "San Diego de los Baños" in Pinar del Río, "Elguea" in Santa Clara, and "Ciego Montero" in Cienfuegos. These are capable of promoting the wellbeing of both the body and mind. The thermal waters, set in paradisiacal landscapes, make you believe that the fountain of youth truly exists. More information about the spas can be found in the article. Cuba's Hot Springs and Sulfurous Waters: Sources of Health and Life, Available at https://onlinetours.es.

The rogations of water, a practice carried out to request rain during times of drought and often associated with Catholic saints, are recognized in many parts of the world. In the eastern province of Santiago de Cuba, droughts have been a chronic problem throughout the city's history. In colonial times, the people of Santiago would pray to Catholic saints for rain. According to Emilio Bacardi, in the year 1673, rogations and processions were held due to a severe drought. "The god of rain weeps over Santiago de Cuba" (Estrada 2021).

Rogativas y cantos para pedir agua

Al Cristo del Amparo
Santo Cristo del Amparo
Nuestro padre y redentor
Agua te pedimos todos
Danos el agua Señor
Danos el agua Señor
Aunque no lo merezcamos

Que si por merecer fuera
Ni aun la tierra en que pisamos
Agua pide el artesiano
Agua pide el labrador
Agua te pedimos todos
Danos el agua Señor

A la virgen de la Asunción

A la Virgen de la Asunción
Agua te pedimos
domingo glorioso
Muy tristes y humildes
Estos tus devotos
Cómo relucen las joyas
señora de tu vestido
así relucía el agua
por los surcos de los trigos
Agua te pedimos
María sin mancha
Porque eres la madre
De toda esperanza
Quién es aquella señora
Que viene por el camino
Es la Virgen la Asunción
Que viene de ver los trigos

Rogatives and chants to ask for water

To the Christ of the Amparo
Holy Christ of the Amparo
Our father and redeemer
Water we all ask of you
Give us the water Lord
Give us the water Lord
Even if we do not deserve it
, for if we did not deserve it,
not even the earth on which we tread
Water the artisan asks for Water
the farmer asks for
Water we all ask of you
Give us the water Lord.

To the Virgin of the Assumption

To the Virgin of the Assumption
Water we ask of you
glorious Sunday
Very sad and humble
These your devotees
How the jewels glitter
lady of your dress
So glistened the water
in the furrows of the wheat.
Water we ask you
Mary without stain
For you are the mother
Of all hope
Who is that lady
Who comes along the road
Is the Virgin the Assumption
Who comes to see the wheat

Conclusions

In Antillean aboriginal mythology, Boinayel held a special significance as a deity associated with rainfall. This belief transcended time and included the curative effects of water and the practice of rogations to procure rain during times of drought.

9.3 Man and Technologies for Water Use

Despite not being abundant what refers to the survival and control of social behavior through mythological expressions in the archeological field, there are studies that contribute to the comprehensive interpretation of the social context of the first settlers who occupied the island (Ortiz 1922; Calvera et al. 2007; González et al. 2019).

The Cuban aborigines covered the whole island always near rivers, lagoons or swampy areas, favorable for aquifer and agricultural exploitation. The aboriginal communities that dominated the territory were found in the mountainous areas and particularly in the areas close to the course of the rivers or where multiple springs poured their waters, which fertilized the soils in a permanent or seasonal way (Fernández et al. 2012).

Studies carried out by (Calvera and García 1994) refer to and support the presence of three habitation patterns: the first one was characterized by isolated habitation sites, as is the case of "Los Buchillones"; a second habitation pattern of agro-pottery sites located on the cusp of hills of not much height and with an abundant watering place, at a short distance, constituted by the sites Santa Clarita, El Güiro and the residuaries located on the margins of the Palma river; and a third settlement pattern for the sites of Santa Clarita, El Güiro and the residuaries located on the margins of the Palma river; El Güiro and the residuaries located on the banks of the Palma River; and a third settlement pattern for the sites located in a marshy lacustrine area near the Loma de Cunagua or Sierra de San Judas de La Cunagua. All these aboriginal settlements had in common the proximity to water sources. After the arrival of the colonizers to the island, the indigenous population dwindled until its total disappearance. It was then that the Spaniards began to bring blacks from Africa as cheap labor.

Several chronicles and articles confirm that when the Spaniards arrived on the island in the fifteenth century, they discovered a pristine land inhabited by aboriginal descendants of the Aruaco stock. These people belonged to a primitive community and had varying levels of development. At the end of the fifteenth century, the indigenous settlements, known as bateyes, spanned across the entire island, including coastal and inland regions, always situated near rivers, lagoons, or swampy areas (Baisre 1987).

During the colonial period in the country (1492–1898), the Spanish colonizers implemented a strategy to populate the island. This strategy, conceived at the beginning of the sixteenth century, involved the establishment of small settlements called "villas" in areas deemed favorable for agricultural and aquifer exploitation. These settlements were accompanied by the distribution or encomiendas of aborigines. The Spanish chose locations that were also favorable for navigation to the lands of the continent, indicating their intention to use the island as a departure site for the conquest of America rather than as a permanent place to stay (Martínez 2013). The emergence of the first water storage systems, such as cisterns or ponds, allowed for the capture and gathering of water from nearby springs, shortening transportation distances for villages and towns.

Also, according to information from researcher Jorge Juárez Cano in Volume I of "Apuntes de Camagüey," this occurred in the sixteenth century. In 2018, he stated that there was a heyday in the production of large tinajones and roof tiles in Cascorro (now a municipality in Camagüey). They were produced in large quantities, primarily for storing drinking water, as is still the case today.

During the seventeenth century, the population of Cuba grew to 10,000 inhabitants and the economy experienced significant growth, largely due to the development of the sugar industry. Numerous sugar mills were constructed, all located near sources of supply. In this century, military constructions have taken a backseat to civil constructions and houses, which typically feature a patio where water is collected in a well or cistern. At this time, rainwater harvesting emerged as an alternative source of water supply. However, due to the expensive nature of excavations in the living rock, it was only implemented in castles and convents (Torres 2019).

In Havana, the capital of the country, the water supply system consisted of the Zanja Real, an 11 km long open canal that supplied various points along its route before reaching the city. During the 17th, 18th, and early nineteenth centuries, the Zanja Real served as the primary source of water for the population, navy, and fleet. It also facilitated the development of irrigated agriculture, including sugar mills, tobacco, wheat, and corn mills. Later, water saws were established on its banks, which ultimately led to a decline in the quality of the water it transported. As a result, a new source of water was needed. In 1835, the Fernando VII Aqueduct was inaugurated. It carried water to Havana inside the walls through cast iron pipes, with the purpose of improving the water supply and hygienic conditions for the benefit of the population. In fact, this aqueduct was a significant advancement in several ways. It included the installation of distribution pipes along the streets, providing direct services to homes, hospitals, barracks, and public fountains. All of these works served as antecedents to the major project that would take over 30 years to construct and would be remembered by the name of its architect: the "Albear Aqueduct." This aqueduct was one of the most remarkable engineering feats of its time on a global scale. Considered one of the seven wonders of Cuban civil engineering, this system collects water from the Vento springs and distributes it through a gravity channel to various reservoirs. The value of this work is immense, and even after 127 years of use, it continues to ensure the water supply for over 15% of the city (Arduengo 2020) (Fig. 9.1).

At the end of the nineteenth century, Cuba transitioned from a colony to a state of dependent capitalism in 1898. The total population at the time was 1.57 million inhabitants, with 65% concentrated in three western provinces: Havana, Matanzas, and Villa Clara. Out of the total population of the island, 31.8% lived in urban areas. Among them, 28.5% resided in cities with more than 20 thousand inhabitants. The country's capital, which had a population of 236 thousand citizens, accounted for 15% of the island's total population.

During the period of the Republic between 1902 and 1958, only a few aqueducts were constructed, and most of them had inadequate water supply sources and lacked the necessary water purification treatments. Notably, the aqueducts built in the major

Fig. 9.1 Albear aqueduct. *Source* Montalván 2020

cities of the former provinces of Las Villas, Camagüey, and Oriente were the most prominent.

Regarding the sewage system, the situation remained critical during this period. It is evident that only partially headwater cities were built. However, it should be noted that since 1959, a group of studies has been developed to obtain observation networks of the hydrological cycle, including rainfall, surface water runoff, sectorization, determination of groundwater potential, and monitoring of their quality (García 2005).

Conclusions

The Cuban aborigines had limited development and settled in areas near rivers, lagoons, and the sea, which facilitated the growth of subsistence activities. The first water storage systems, such as cisterns or ponds, were built in the fifteenth century. The production of "tinajones" was encouraged during the sixteenth century to store liquids during times of scarcity. In the seventeenth century, rainwater collection became a popular alternative source of supply, often stored in wells or cisterns. In the nineteenth century, the Albear aqueduct was inaugurated. It allows for the collection of water from springs and distributes it through a gravity channel to various reservoirs. This system has been in use until the present day. Aqueducts in the country were insufficient at the beginning of the twentieth century.

9.4 Engineering Works

According to Batista (2019), there are a total of 243 constructed reservoirs and hundreds of smaller ones, known as micro-dams, scattered throughout the region. Their purpose is to supply water to various users. However, there are areas where rainfall is scarce, while in other regions, precipitation exceeds the average annual values. In Cuba, the primary source of water supply is rainfall. As a result, the country's socioeconomic development is directly dependent on this climatic factor. For decades, there has been a consideration to transfer water from basins with abundant water resources to those with fewer resources. To learn more about water transfers in the Cuban archipelago, we recommend reading the article "Los trasvases en el archipiélago cubano" published in Revista Voluntad Hidráulica No. 129.

There are currently ten water diversion systems in operation, each with its own unique operating regime. This hydraulic infrastructure includes approximately 292 km of canals, conduits, and tunnels, which enable the transfer of 262 million cubic meters of water annually. Of these, three stand out as the largest in the country, and they are still under construction. However, they already benefit more than 52,600 hectares of agricultural land (Yilén 2020).

9.4.1 Center-East Transfer

The water transfer system, which dates back to the 1970s and 1980s, is the oldest of its kind. It spans 315 km and boasts impressive levels of execution. The system includes 85 km of constructed canals, six dams (Zaza, Iguanojo, Banao II, Lebrije, Dignorah, and La Felicidad), and two diversion dams (El Patio and Sur del Jíbaro). The area under irrigation in Sancti Spiritus is approximately 47,000 hectares, making it more valuable in terms of use. Of this, 38,000 hectares are used for rice cultivation, various crops, and sugar cane. In Ciego de Avila, 9000 hectares of land are irrigated, benefiting the cultivation of sugar cane, citrus fruits, grains, vegetables, and root crops. There are 212 fish breeding ponds, with 200 currently in operation in the province of Espiritu and 5 out of 28 in operation in the province of Avila. According to data from the National Institute of Hydraulic Resources (INRH), it generates 41.5 GWh and saves 10,794 tons of fuel. Priorities are focused on actions that increase the utility value and modify the irrigation system. Work is also being done on the infiltration channel, which will allow for loading of the water table if necessary, as well as the construction of regulating reservoirs. The lining of the Zaza Canal in Sancti Spíritus province is also undergoing rehabilitation to ensure a safe water supply to the fertile lands of Ciego de Avila and Camagüey (Fig. 9.2).

Fig. 9.2 Central-East diversion: a river crossed. Photograph Borrego 2012 (Coordinates not provided)

Fig. 9.3 North–South transfer. Photograph Valera 2017 (Coordinates not provided)

9.4.2 North–South Transfer

The construction of a diversion in the Sabanalamar River is the solution to the lack of water for agriculture in the Caujerí Valley, located east of the city of Guantánamo. The diverted water will flow through a tunnel by gravity and reach Pozo Azul. From there, it is also delivered to the farmers by gravity. It can transfer 1 m^3/s of water, and during periods of heavy rainfall, it can transfer up to 5 m^3/s. It replaces costly pumping, which used to cost approximately US$ 700,000 at fuel prices (Fig. 9.3).

9.4.3 East–West Transfer

It consists of six stages: two have been completed, one is currently in progress, and three are in the project and planning phases. In the first phase, the water supply to Holguín and the tourist area in the northern part of the province was ensured. An area of 4179 hectares has been brought under irrigation, and since 2011, floods have been prevented through the construction of a spillway. Currently, the third stage is underway with the construction of a 15 km canal between Birán and Báguano in the province of Holguín. It is also possible to synchronize the implementation of the water transfer with the investments of the Agriculture and Sugar Company,

Fig. 9.4 East–West transfer. Photo Leyva 2020 (Coordinates not provided)

which plans to irrigate around 700 hectares of land. In the context of rice cultivation, the Nipe Deleite-Cosme Herrera canal was constructed to ensure a consistent water supply for 2000 to 3000 hectares of rice fields (Fig. 9.4).

Conclusions

Three water diversions stand out as the largest in the country: the Central-East, North–South, and East–West water diversions. These diversions ensure a reliable water supply for the population, irrigation in priority crop areas, energy generation, and fuel savings.

9.5 Water, Sanitation, and Challenges

Before 1959, Cuba had 300 settlements with over 1000 inhabitants, but only 114 of them had aqueducts and only 12 had sewage systems. At that time, the island had 16 chlorination facilities, 4 water treatment plants, and a treatment plant located in Santa Clara. After 1959, which coincided with the triumph of the Cuban Revolution, measures were taken to reverse this situation, resulting in a considerable increase. By the end of the twentieth century, there were more than 3000 aqueducts built. In 2000, a process of reorganization began for the INRH, during which its structure, functions, and organizational responsibilities were modified. This led to the emergence of companies and business groups that implemented a new system of management and business administration (García 2007a).

However, the lack of wastewater treatment continues to be one of the sanitary problems to be faced, mainly in third party providers where it is still considered low. Similarly, there are approximately 100,000 septic tanks in an unfavorable state of construction, with over 37,000 requiring frequent cleaning. This not only affects the cleaning cycle and increases energy consumption, but also impacts sanitation coverage as these waste products are generally not treated (INRH 2019).

As referenced by Cuellar et al. (2017), a surveillance and control system is currently in place to ensure that households receive water of the necessary quality. Research is conducted to support this surveillance system, allowing for the identification of areas with greater sanitary risks and limited access to drinking water. This research also analyzes the potential consequences for the health of the population and studies diseases caused by water contamination, among other aspects aimed at improving the quality of life of the population and enhancing the availability of water and sanitation services with effectiveness. Treatment, reinforced by systematic monitoring, contributes to optimizing the sanitary quality of water for human consumption and reducing the health risks associated with diseases caused by water contamination.

In fact, in an interview, Rodriguez (2018) made reference to the fact that INRH has identified challenges that will undoubtedly allow for a decrease in water consumption, and that the efficiency of the system can be measured.

- Challenge to Address Low Hydrometric Coverage.
- The extended national water program until 2030 includes projects for water supply, sanitation, storage, transfer, and hydrometric works. It also includes investments in necessary equipment and maintenance.
- Construction of seawater desalination plants. These plans include not only provisions for supplying the population, but also everything necessary for agriculture, hotel infrastructure, and the housing program.
- It is planned to redesign the observation network for both groundwater and surface water quality.
- It is proposed to achieve, by 2030, an energy matrix in which 24% of electricity generation comes from renewable sources, to reduce the use of fossil fuels by half, and to increase the forested area of the national territory by two percentage points.
- In the context of greenhouse gas (GHG) mitigation, priority is given to sectors such as construction, agriculture, forestry, and electric energy. A particular strategy is being promoted to capture all emissions in the swine sector, which is a rapidly growing industry in our economy and a significant source of residual pollutants.

Conclusions

There was an increase in the number of aqueducts and sewage systems at the end of the twentieth century compared to previous periods. However, the lack of wastewater treatment remains a problem that needs to be addressed. Fortunately, the implementation of monitoring and control systems ensures that water reaches homes with the necessary quality. The expansion of hydrometric coverage, the elimination of leaks, and the construction of desalination plants will decrease water consumption.

9.6 Impacts of Climate Change

In particular, the Small Island Developing States (SIDS) in the Caribbean share many similarities that increase their vulnerability and reduce their resilience to climate variability and change (CC). These similarities include their small land area, remoteness, population growth, susceptibility to natural disasters and extreme weather events, extreme openness of their economies, and limited capacities and resources (IHP 2015).

Various authors have reported that Caribbean countries and Small Island Developing States (SIDS) are facing significant water security challenges, including water scarcity, drought, pollution, watershed degradation, saline intrusion, and flooding. An anthropogenic impact is evident in the heightened occurrence of multivariate extremes, including droughts (Diffenbaugh et al. 2015) and wildfires (Abatzoglou and Williams 2016; Radeloff et al. 2018).

Several publications by different authors have reported alterations in the precipitation regime and the effects of droughts in the Caribbean region (Ocha 2015; Blunden and Arndt 2016). Additionally, there has been an increase in temperature and potential evapotranspiration (PET) (Ocha 2015; Blunden and Arndt 2016; Herrera and Ault 2017). However, it is unclear whether these changes are a result of anthropogenic warming and whether their severity has increased.

In recent years, significant changes in the climate have been observed. The most notable evidence is the rise in mean annual temperature, which is attributed to an increase in minimum temperature. Additionally, there has been a decrease in cloudiness, more intense and prolonged droughts (although less frequent), an increase in precipitation exceeding 50 mm, and a greater anticyclonic influence (Boudet 2015).

Recent research on global model predictions for the region and the downscaling produced by regional climate models (RCMs) indicates that temperature increases are projected for the entire region. These increases are generally lower than global increases and their values depend on the assumed scenario and subregion, with a range of variation between 2 and 6 °C. Projections of precipitation variation show a general downward trend, although with a strong subregional dependence. Therefore, it is important to pay attention to the decrease in water potential, which is a result of the impact of climate change and is manifested in a progressive reduction in water availability (Vichot ct al. 2019).

According to Planos (2022), the country's current potential water resources are estimated to be 33 km^3, which is 15% less than the previously estimated 38.1 km^3. By the year 2030, it is projected that the potential water resources will decrease to 30 km^3, resulting in a deficit of over 21% compared to the figure endorsed by INRH. He asserts that the current and future availability of usable water is becoming increasingly critical. Not all potential water resources are usable, and groundwater resources are also at risk due to the impact of marine intrusion and foreseeable changes that will intensify natural recharge and the static level of aquifers. The spatial distribution of the runoff deficit by 2030 indicates that, regardless of the future climate scenarios, there will be a reduction in water availability throughout the country. This reduction will be more pronounced in the western and eastern regions, although the regional differences are not significant. After considering these factors, it is essential to engage governors, decision-makers, researchers, specialists, and technicians to continue working on effective water resource management.

Conclusions

Small island states are facing significant water security challenges, including water scarcity, drought, pollution, watershed degradation, saline intrusion, and flooding. These issues are exacerbated by changes in precipitation patterns, rising temperatures, and increased potential evapotranspiration. Cuba has estimated a decrease in usable water resources for all future climate scenarios.

Annex 1 Acronyms

Aljibe, cistern or tinaco: used to store drinking water for consumption by the population.

Bateyes: name given to villages located in the middle of sugar cane plantations.

Ponds: small cavity of water, natural or artificial, used daily for irrigation, fish breeding, swimming, etcetera.

Sugar mills: the set of industrial facilities dedicated to the milling and processing of sugarcane to produce sugar or sucrose.

Pozos or pozas: pond; concavity in which there is detained water. Well of a river, place where the river is deepest.

Tejar: place where tiles and bricks are manufactured.

Tinajón, tinajas: earthenware container used to store water.

Diversions: aims to supply water from one basin to another.

Urbe: city, especially one with a large number of inhabitants.

Villas: town that in the Middle Ages possessed certain privileges that distinguished it from other places.

Vessel: vessel capable of holding something: liquids such as water or oil; semi-liquids.

Water will: government's coherent response to ensure water quantity and quality for economic and social development and environmental protection in the country.

Annex 2

See Figs. 9.5, 9.6 and 9.7.

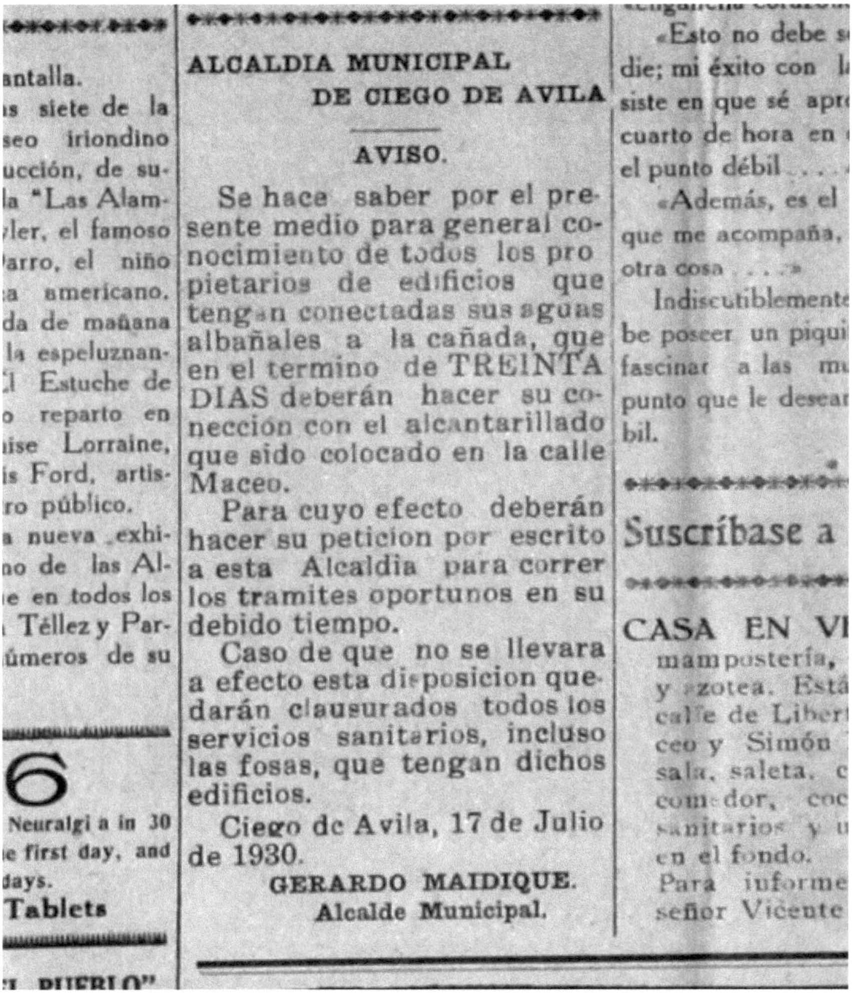

ALCALDIA MUNICIPAL
DE CIEGO DE AVILA

AVISO.

Se hace saber por el presente medio para general conocimiento de todos los propietarios de edificios que tengan conectadas sus aguas albañales a la cañada, que en el termino de TREINTA DIAS deberán hacer su conección con el alcantarillado que sido colocado en la calle Maceo.

Para cuyo efecto deberán hacer su peticion por escrito a esta Alcaldia para correr los tramites oportunos en su debido tiempo.

Caso de que no se llevara a efecto esta disposicion quedarán clausurados todos los servicios sanitarios, incluso las fosas, que tengan dichos edificios.

Ciego de Avila, 17 de Julio de 1930.

GERARDO MAIDIQUE.
Alcalde Municipal.

Fig. 9.5 Notice to the population about connection to the sewerage system. El Pueblo newspaper (1930, July 17)

Fig. 9.6 News about connection of houses to sewerage and water prices. El Pueblo newspaper (1930, August 20)

experimentando depre
en el precio de ese artículo,
o es casi una necesidad
sis, por las grandes canti-
ue a bajo precio pueden
rse de otros países que,
l Brasil, tienen considerable
en su producción; la reso-
residencial a que dejamos
eferencia, tiene como punto
l, el elevar los derechos
rios a todo el café que
el país a más de los
cincuenta mil quintales que
ara equilibrar lo que pro-
, con lo que reclaman las
ades del consumo; fijándole
e se importe con exceso
ciento cincuenta mil quin-
e necesitamos para igualar
ucción con el consumo, la
áxima señalada en el aran-
representa el doble de lo
ctualmente paga, todo el
e en el país entra, con en-
completa libertad.
eso se reduce lo que pre-
los cafetaleros, creemos que
e demorarse el concedérse-
que piden; eso sí estudiando
bien todos los cálculos, com-
dose debidamente cual es

Venga, pués lo del vivero o
cualquier otra solución clara y pronta que evite el "vivio".

Entre los titulares de "El Camagüeyano", de ayer, aparece el siguiente: "Fueron autorizadas 5 vaquerías declaradas en buenas condiciones para la venta de leche cruda".

Y a renglón seguido:

"Quedó ayer instalada la planta provisional para depuración de las aguas del alcantarillado".

Si esas noticias tienen alguna relación entre sí, tenemos que suponer que en la Ciudad prócer, la escasez del agua para las necesidades del consumo sigue constituyendo un problema; por lo que al aumentar tales necesidades, impuesto por la "cristianización" del jugo lácteo, hay que recurrir a lo de la depuración del agua que ya fuera utilizada, hasta en los más bajos menesteres.

Nos parece que esto de la leche, todavía no se ha arreglado en la Capital de nuestra Provincia.

YOYO.

No quiera usted mencio
todas sus mercancías en un
lo anuncio.
No diga usted en sus an
cios lo que no es cierto.
No espere usted que un
anuncio convenza a todos.
No crea usted que debe d
de anunciar porque sus n
cios no vayan del todo bien
No anuncie usted las difi
tades de su negocio.
No riña usted con el com
tidor, ni mucho menos co
cliente.
No deje de investigar
razón y anunciarla.

Andrea Mor
COMADRO
Máximo Gómez 28

Colegi
Primera y S
Comercio - Taq

Fig. 9.7 Notice to the population about the installation of a sewage treatment plant. El Pueblo newspaper (1930, August 20)

References

Abatzoglou JT, Williams AP (2016) Impact of anthropogenic climate change on wildfire across western US forests. Proc Natl Acad Sci USA 113:1177011775. https://doi.org/10.1073/pnas.1607171113

Arduengo DA (2020) The Fernando VII aqueduct: an ignored revolution in water management in Havana. Cuban J Ciencias Técnicas V.10(1). Available at: www.revistaccuba.cu/index.php/revacc/article/view/755/776

Arrom JJ (1990) Relación acerca de las antigüedades de los indios. Fray Ramón Pané. Editorial Ciencias Sociales, Havana, Cuba. Available: https://www.dlls.univr.it/documenti/Avviso/all/all287650.pdf

Ashfaq M, Cavazos T, Reboita MS (2020) Robust late twentyfirst century shift in the regional monsoons in RegCMCORDEX simulations. Clim Dyn. https://doi.org/10.1007/s00382-020-05306-2

Baisre (1987) La pesca en Cuba: apuntes para su historia (I): la pesca en las comunidades primitivas: etapa precolombina. Revista Mar y Pesca 262:38–39. http://hdl.handle.net/1834/5053

Batista Silva JL (2019) Los trasvases en el archipiélago Cubano. Revista Voluntad Hidráulica 129: 41–47. Havana, Cuba, ISSN 0505-9461

Blunden J, Arndt DS (2016) State of the climate in 2015. Bulletin of the American Meteorological Society V 97(8):SiS275. https://doi.org/10.1175/2016BAMSStateoftheClimate.1

Boudet D (2015) Impacts of climate change in Cuba. Regional workshop on adaptation to climate change at the local level. Spanish cooperation training center in Cartagena de Indias. Available at: https://cambioclimatico-regatta.org

Calvera J, García A (1994) A palustrine lacustrine system as a habitat for agro-palustrine communities. Archaeological studies. Compilation of themes. Editorial Academia. Havana. pp 1–6. https://www.eumed.net/rev/caribe/2019/02/asentamiento-aborigen-cuba.html

Calvera J, Valcárcel R, Ordúñez (2007) Wood in the archaeological world of Los Buchillones. Office of the Historian of the City of Havana. Havana, no. 6

CITMA (2017) Plan Nacional de Enfrentamiento al Cambio Climático. Available at: www.citma.gob.cu/. Accessed 5 Apr 2021

Cuellar L, Maldonado G, Cepeda Y (2017) Quality of water for human consumption. Cuban J Hygiene and Epidemiol 55(1):13–18. ISSN 1561-3003, Available at: http://www.revepidemiologia.sld.cu/index.php/hie/article/view/75/149

De las Cuevas (2001) 500 years of constructions in Cuba. Chavín Servicios Gráficos y Editoriales, S.I., Madrid

Diffenbaugh NS, Swain DL, Touma D (2015) Anthropogenic warming has increased drought risk in California. PNAS 112(13):3931–3936. First Published 2 Mar 2015. https://doi.org/10.1073/pnas.1422385112

Estrada J (2021) The god of rain cries over Santiago de Cuba. Gaspar, the local, 5 Aug. http://www.ellugareno.com/2021/08/el-dios-de-la-lluvia-llora-sobre.html

Fernández R, González J (2001a) The enigma of the aboriginal petroglyphs of Cuba and the Insular Caribbean. Editorial Centro de Investigaciones y Desarrollo de la Cultura Cubana Juan Marinello, Havana

Fernández R, Morales D, Torres La Paz L (2012) Water-related cosmovision in farming groups, its symbolism and meaning: Báguano region, Holguín, Cuba. Ciencia Ergo Sum 19(2):145–155. Autonomous University of the State of Mexico. ISSN: 1405-0269. Available: https://www.redalyc.org/pdf/104/10422928005.pdf

García R (2005) In: Hydraulics in the Cuban revolution. Institute of History of Cuba

García R (2007a) La Hidráulica en Cuba: experiencias y resultados (1959–2005). In: CD I Workshop use and care of water, Havana, November 16–17, 2006, Edited by the Economic Society of Friends of the Country, Havana

González UM, Rodríguez R, Reyes I, Yero JM, Chinique Y, Roksandic M (2019) Horadada lithic discs in aboriginal archaeological contexts of Cuba. Approach to their social function from ethnography and archaeology. Sci Soc V 44(4):7–23. https://doi.org/10.22206/cys.2019.v44i4.pp7-23

González HUM (2015) Las crónicas generales de Indias en la Arqueología de Cuba. Limits and perspectives in the ethnohistorical reconstruction of aboriginal societies. Cuba Arqueológica. Digital J Archaeol Cuba and the Caribbean. 4–80. Havana, Cuba, Colombian Institute of Hispanic Culture, Human Geography of Colombia. ISSN: 1852-0723

Herrera D, Ault T (2017) Insights from a new high-resolution drought atlas for the Caribbean spanning 1950 to 2016. J Clim 30(19):7801–7825. https://doi.org/10.1175/JCLI-D-16-0838.1

IHP (2015) Graphic groundwater and climate change small Island developing states. Potential effects of climate change on groundwater in SIDS. pp 5. Available: https://groundwaterportal.net/sites/default/files/GRAPHIC%20-%20Aguas%20Subterraneas%20y%20Cambio%20Climatico%20PEID.pdf

INRH (2019) Integration to INRH's work objectives, the goals of the national environmental strategy, sustainable development goals (SDGs) and the climate change agreement

Jiménez A, Cortobius M, Kjellén M (2014) Water, sanitation and hygiene and indigenous peoples: a review of the literature. Water Int 39(3):277–293. https://doi.org/10.1080/02508060.2014.903453

Jiménez A (1981) Taino myths. Transformación de niños en tonas o animales a manera de ranas, Listín Diario supplement. Saturday 5 Dec, Santo Domingo

Martínez F (2013) The road to a different city. Urbanism as a discipline. Emergence and evolution of human settlements in Cuba. Revista planificación física cuba. No. 16. ISSN 0138-6239. www.ipf.gob.cu/es/revista/revista-planificación-física-cuba-3

Ortiz F (1922) History of Indo-Cuban archaeology. Havana, Imprenta "El siglo XX"

Paz L (2010) Climate change and its consequences for Cuba. Revista Bimestre Cubano de la Sociedad Económica de Amigos del País, January-June. Epoch III, vol 107(32). Havana, pp 54–71. Available from: https://www.researchgate.net/publication/325106326-Cambio-climaticoy-medio_ambiente_en_Cuba_Formulacion_de_politicas_y_medidas_deadadadaptacion_para_el_bienestar_humano_y_funcionamiento_de_la_sociedad_Climatic_change_and_environment_in_Cuba_Formulation. Accessed 22 Nov 2021

Planos E (2022) Approximation to the potential water resources in Cuba to 2030. In: Hydraulic and environmental engineering, vol 43(1). pp 48–62

Radeloff VC, Helmers DP, Kramer HA, Mockrinb MH, Alexandre PM, Massada AB, Bustic V, Hawbaker TJ, Martinuzzi S, Syphard AD, Stewart SI (2018) Rapid growth of the US wildland-urban interface raises wildfire risk. NAS 115(13):3314–3319. https://www.ncbi.nlm.nih.gov/pmc/articles/PMC5879688

Rodríguez A (2018) Intervention of the delegation of the Republic of Cuba to the United Nations, at the high-level event international decade for action water for sustainable development (2018–2028), 26 Mar 2018. http://misiones.minrex.gob.cu/es/articulo/cuba-da-la-bienvenida-al-decenio-internacional-para-la-accion-el-agua-para-el-desarrollo

Sanchez D (2018) Water, lots of water in hell, 6 Feb, Cubadebate. http://www.cubadebate.cu/fotorreportajes/2018/02/06/agua-mucha-agua-en-el-infierno/

Torres R (2019) Rainwater harvesting as a solution in the past and present. Riha 40(2) Havana May-Aug

Vichot AL, Benzanilla A, Martínez D, Centella A (2019) Current status of the application of downscaling methods to climate change projections in Central America and the Caribbean. Cuban J Meteorol 25(2). E-ISSN 0864-151X

Yilen C (2020). Behind the route of the water transfers in Cuba. Cubahora Revista digital. https://www.cubahora.cu/sociedad/tras-la-ruta-de-los-trasvases-en-cuba-graficos

Chapter 10
Evolution of Approaches on Water Resources Management and Integrated Watershed Management Planning: Complement, Synergies, and the Way Forward

L. Benegas-Negri

Abstract Several approaches were used since the beginning of times to do water management. Relatively recent is the integrated water resources management (IWRM) proposed as an intent to bring a standard approach which is currently included in all recent water laws in Latin America and the Caribbean. An older approach, also guiding water and natural resource management is integrated watershed management (IWM). The beginning of this concept in Latin America probably started in the 1950s, focused on planning and building infrastructure for water exploitation. An evolution toward a conservationist approach without focusing on the human being occurred in the 1980s. In the 1990s the focus changed toward the family, people, and communities, but still lacking participation in decision-making processes. These stages of IWM are sectoral, usually under the exclusive competence of governmental agencies. Around 2006 to date, IWM has an integral systemic and interdisciplinary approach; understood as co-management. Nowadays, emerging concepts as nature-based solutions, ecosystem-based adaptation, resilience of socioecological system, and one water reinforce and guide adjustments in watershed and water resource management practice. Such emerging concepts, together with the sustainable development goals, properly integrated, could visualize how IWRM and IWM are the best vehicles to operationalize all of them.

Keywords Water-related approaches · Planning · Model of co-management of watersheds · Concertation · Common agenda

L. Benegas-Negri (✉)
Campus del CATIE, 30501 Turrialba, Cartago, Costa Rica
e-mail: laura.benegas@catie.ac.cr

G. A. Corzo Perez et al. (eds.), *Aqua Vitae*, The Latin American Studies Book Series,
https://doi.org/10.1007/978-3-031-57731-4_10

10.1 Evolution of the Watershed Management Concept

To summarize the evolution of watershed management concept, first it is necessary to define the term watershed. Physically a watershed is the area that drains to a common outlet. It is the basic building block for land and water planning (Darghouth et al. 2008). Expanding this definition, we can consider the watershed as a hydrological response unit, a biophysical unit, as well as a holistic ecosystem in terms of the materials, energy, and information that flow through it (Wang et al. 2016), although the understanding of the holistic and ecosystemic approaches made more complex to work within and in the context of watersheds.

The beginning of the watershed management concept in Latin America probably started in the 1950s, when Mexico created executive commissions for planning and building infrastructure projects or for the study of water resource exploitation in a framework of regional economic development with water as the integrative territorial resource (Dourojeanni et al. 2002; Ozyuvaci et al. 1997).

There was an evolution toward a conservationist approach in the 1980s, with the aim of assuring water quality, reforestation, or soil conservation without focusing on the human being but in the natural resources, and where people's participation was limited (Darghouth et al. 2008).

In the 1990s, the focus changed toward the family, people, and communities. In this approach human beings are considered as allies for conservation, protection, and reasonable use of natural resources, aiming equilibrium of production with conservation. Nevertheless, participation of the population in the decision-making processes was still limited (Darghouth et al. 2008).

The above-mentioned stages of watershed management are sectorial approaches, under the exclusive competence of governmental agencies. However, since approximately 2006 to date, watershed management evolved toward an integral, systemic, and interdisciplinary approach, which is understood as participatory, join, and integrated management or co-management (Faustino and Jimenez 2005; FAO 2007; Benegas and Faustino 2008; Kammerbauer et al. 2010).

Emerging concepts as ecosystem services, community and ecosystem-based adaptation for risk management, and resilience and vulnerability to stress and shocks as climate change are some of the approaches which have been helping to adjust the watershed management approach, although the planning focus as a valid tool is still demonstrated, and recently there are different "last names" for the watershed management approach. For example, the recent water security planning from USAID-SWP (2016) was presented as a tool kit describing step by step how to build such water security plans.

It is important to briefly address the meaning of each of the key emerging approaches as follows. Ecosystem services place the environment as natural capital, one of society's important assets. But ecosystem services are becoming increasingly scarce, and the approach of ecosystem services search to provoke society to acknowledge the value of natural capital. Here, ecosystem services valuation (ESV) is the method to tackle such a challenge (Liu et al. 2010).

Another relevant concept is resilience, meaning the ability of a socioecological system to undergo change and retain sufficient functionality to continue to support livelihoods through, for example, the sustained provision of ecosystem services related to food security, access, and use of such resources (Park et al. 2010). The concept of socioecological systems highlights that people and nature are inextricably linked and function within complex adaptive systems (CGIAR Research Program on Water, Land and Ecosystems-WLE 2014). Closely related to resilience, the approach of community-based adaptation emerges, focusing on adaptation as development (Forsyth 2013). Here adaptation is local and place-based; it addresses the locally and contextually specified nature of climate change vulnerability because it takes place at local levels where people encounter impacts, build adaptive capacity, and respond to shocks. With this approach, adaptation strategies must be generated through participatory processes, involving local stakeholders, and developing disaster risk reduction practitioners, rather than being restricted to impacts-based scientific inputs alone (Ayers and Forsyth 2009). Vulnerability encompasses a variety of concepts and elements including sensitivity or susceptibility to harm and lack of capacity to cope and adapt. Sensitivity is the degree to which a system or species is affected, either adversely or beneficially, by climate variability or change (IPCC 2007). The effect may be direct (e.g., a change in crop yield in response to a change in the mean, range, or variability of temperature) or indirect (e.g., damages caused by an increase in the frequency of coastal flooding due to sea level rise). Exposure is the presence of people, livelihoods, species or ecosystems, environmental functions, services, and resources, infrastructure, or economic, social, or cultural assets in places and settings that could be adversely affected; and to complete the concept of vulnerability, one must consider the adaptive capacity as the ability of systems, institutions, humans, and other organisms to adjust to potential damage, to take advantage of opportunities, or to respond to consequences (IPCC 2007; Winer et al. 2020).

The newest and more global concept regarding natural resources is nature-based solutions (NbS) defined by IUCN as "actions to protect, sustainably manage, and restore natural or modified ecosystems, that address societal challenges effectively and adaptively, simultaneously providing human well-being and biodiversity benefit" (Cohen-Shacham et al. 2016).

10.2 Integrated Water Resource Management

Integrated water resources management (IWRM) is an approach that helps to balance competing water demands from across society and the economy, without compromising the sustainability of vital ecosystems. This is achieved through coordinated policy and regulatory frameworks, management arrangements, and financing (GWP 2000; UNEP 2021). IWRM was defined in 2002, at the Johannesburg World Summit on Sustainable Development (WSSD) by the Technical Advisory Committee of the Global Water Partnership (Rahaman and Varis 2005).

IWRM is structured around four dimensions (UNEP 2021): 1. Enabling environment, where national and subnational policies and laws outline the importance of integrated approaches to water resources management. Plans are needed to operationalize policy and regulatory frameworks. 2. Institutions and participation, here institutions and stakeholder participation across sectors are needed at all levels to implement plans and enforce regulations. 3. Management instruments, where data and information need to be provided to all relevant stakeholders for informed decision-making, covering aspects such as sustainable use, pollution control, ecosystem management, and disaster risk reduction. 4. Financing, here budgets at the national and local level, for investments and ongoing infrastructure and management costs, are needed to implement management instruments and fund institutions. Revenue raising is an important part of this dimension.

10.2.1 The "Schools" of Integrated Watershed Management (IWM) and Integrated Water Resource Management (IWRM)

Both "schools" are very similar and complement each other. However, there are some discrepancies between both approaches. Some experts state that integrated water resources management is the broader umbrella including watershed management, considering all water-related sectors and interested parties. In this case, IWRM "school" implies that watershed concept is neither universally applicable nor comprehensive enough to incorporate all the necessary considerations for effective water resources management (Cardwell et al. 2006). Another group of experts indicate that water is originated from watersheds, and its management covers a broader range of water-related issues along with other influencing factors on water cycle. Thus, IWM is considered an umbrella under which other features of IWRM, such as participation is integrated (Cohen and Dadvison 2011).

The watershed scale also contains all of water resources according to "ridge to reef" concept (Guiang et al. 2021) which seeks the interconnections between the natural and social systems from the "ridges," through coastal watersheds and habitats, and across coastal water bodies (i.e., lagoons) to the bordering "reef" environments (GEF 2016). All stakeholders live in watersheds, and every point in the planet lies in a watershed (Water Science School 2019).

A possible differentiation in these concepts may be interpreted as follows: IWM is an adaptive, integrated, multidisciplinary, participatory systems approach to management of all the resources in a watershed that aims to preserve productivity and ecosystem integrity defined from the water as the resource which integrate the landscape, with soil, plants, and animals within a watershed, thereby protecting and restoring ecosystem services for environmental, social, and economic benefit.

And IWRM is a systematic process for allocation and monitoring of water resource use in the context of social, economic, and environmental objectives, which according to Conca (2006) became the rallying point for international water policy. But, in 2021, UNEP reported that, although the global call for IWRM implementation was formalized in 1992, almost 30 years later, 87 countries (−47%) of their assessment-universe still report "low" or "medium–low" levels of IWRM implementation (UNEP 2021).

Besides such differences and similarities, both approaches need to be used and adapted to the context at hand, given that both constitute the vehicle for water security and overall sustainability of natural resources.

10.3 Watershed Management Plans, Why Are They Still Important?

When we understand watershed management as a systemic approach, the natural relevance of this way of planning is easily justified. According to Jimenez (2005), this systemic approach means that the watershed "is a whole," it cannot be divided in terms of its functions, but at the same time, it is interdependent. In a watershed system, we must recognize several subsystems, i.e., social, cultural, economic, political, legal, technologic, productive, biologic, and physical. The systemic approach for watershed management also implies interactions and interconnections between the upper, middle, and lower or coastal parts of the watersheds, where watersheds management actions must be based in participatory processes to detect causes and consequences of the main problems present in the watersheds, as well as recognizing their opportunities and advantages derived from a rational use of watershed's potential. Also related with a system, Leon et al. (2012) propose to consider the watershed as an integrated system of water fluxes where it can be promoted the territorial and institutional ordering, starting from the common interests.

10.4 The Experience of the Tropical Agricultural Research and Higher Education Center (CATIE)'s Approach of Watershed Planning and Management

The watershed management plan is a tool, which guides the territorial management of the well-known exclusive planning unit for water resource management: the watershed.

The general goal of a watershed management plan is to assure that a watershed's population develop, assess, and carry out processes toward the sustainability of natural resources (water, soil, and biodiversity), allowing productive activities in harmony with the environmental services.

A watershed management plan sets the overall objectives of watershed management interventions and defines the land and water management policies, institutions, and investments needed to achieve the objectives (Qi and Altinakar 2013; Guo 2014; Jimenez y Benegas 2019).

A watershed management program operationalizes the plan by defining the specific policy and institutional changes and investments to be made and the areas to which they will be applied within a specific period. With such a program, it is possible to quantify and calculate the costs of inputs, outputs, and outcomes for watershed management.

A watershed management project is an investment operation within the program, usually over 5–7 years, with specific indicators of performance (Faustino y Jimenez 2005). This investment analysis is strongly suggested because investments designed both to achieve sustainable natural resource management and to improve livelihoods through intensification of resource use are at the heart of watershed management programs.

All actions implemented through watershed plans could start to respond to the following basic reasons which watershed management should pursue (Table 10.1).

Since almost 30 years, CATIE has been working in watershed management. In recent years, this institution has evolved and promoted the adaptive and co-management models in watersheds for Central American region (Kammerbauer et al. 2010). Figure 10.1 shows the CATIE's model of watershed co-management, based on the experiences in four subwatersheds, located in Honduras and Nicaragua, used as practical laboratories or models to test such co-management proposed (Castellón and Prins 2009; Kammerbauer et al. 2009; Lara et al. 2007; Reyes et al. 2008). The main objectives of the model are a joint vision among the stakeholders present in the basin, the appropriation of collaborative objectives, the definition of duties and responsibilities of the stakeholders, and the prevalence of collective over private interests (Kammerbauer et al. 2009).

The model visualizes watersheds as integral systems of water flows of public and collective interest (Kammerbauer et al. 2010). The central axis of the co-management model is water as an integrating resource and the watershed as a planning unit. The basic purpose is to ensure the quantity and quality of water, acting by means of concerted platforms and dialogue among local stakeholders to develop a common territorial agenda, where efforts and resources complement each other. This is based on the existence of groups of interests, mainly local authorities, grassroots organizations, national organizations with a local presence, private enterprise, and development organizations (Jimenez and Benegas 2019).

The model seeks the constitution, installation, and operation of formal and informal spaces and dialogue platforms or the strengthening of existing spaces. The platform acts as an adjunct body and space for convergence, articulation, concertation, and coordination of local organizations and stakeholders, leading to the development of co-management plans to direct, guide, and facilitate decision-making related to the management of the watershed and its natural resources (Kammerbauer et al. 2009).

The model requires the formulation of a vision and a common agenda (León et al. 2012) in the watershed as a co-management plan, which should be inclusive and must

Table 10.1 Main aspects and features supporting watershed management planning

Feature	Reason to use watershed management planning
Functionality	The watershed is a functional region established by physical relations, is more than the sum of parties, and interacts within its subsystems and around its context
Biological linkages	The watershed approach is logical for evaluating biological linkages of upland and downstream activities because within the watershed they are linked through the hydrologic cycle
Holistic	Enables planners and managers to consider many facets (subsystems) of resource development
Environmental impact toward ecosystem services approach	Land use activities and upland disturbances often result in a change of environmental impacts that can be readily examined at the watershed in terms of ecosystem goods and services
Human interactions impacting the social-ecological system of the watersheds	The watershed provides a framework to analyze how the primary decision-makers in the basic unit of planning: "the farms, the private or communal properties or lands" provoke changes in the environment. The environmental impact within the watershed operates as feedback loops for changes in the social-ecological system, which is a way to represent the watershed
Economic logic linked with ecosystem services approach	Many of the externalities involved with alternative land management practices on an individual farm are internalized when the watershed is managed as a unit. Nevertheless, such externalities need to be assumed by sectors and actors who demand ecosystem services from the individual farms. An exercise of prioritization of good watershed management practices (Benegas et al. 2017) could be useful to define mechanisms of compensation or incentives to the stakeholders which maintain ecosystems services in the watershed
Adaptive management to be part of programs and projects	In general, there are policies, government plan, and different national and international mechanisms to finance watershed management practices. Examples of those programs include forestry, soil conservation, rural community development, faming systems, and more recently climate funds to accomplish the Paris agreement (UN 2016), resilience-related projects from international cooperation, actions to accomplish the sustainable development goals, among others, compromise of the countries

Source Adapted from Dixon and Easter (1991), Jimenez (2005), Jimenez and Benegas (2019)

represent the interests of all stakeholders. To implement actions in the privileged territories, the model includes the creation, establishment, and operationalization of an environmental fund as a financing mechanism of priority actions for the management of the watershed (Kammerbauer et al. 2010; Jimenez y Benegas 2019). To design and implement the common agenda, the model developed a set of strategies and instruments for territorial management practices in the critical zones where water is retained for human consumption and provision of ecosystem services, known as

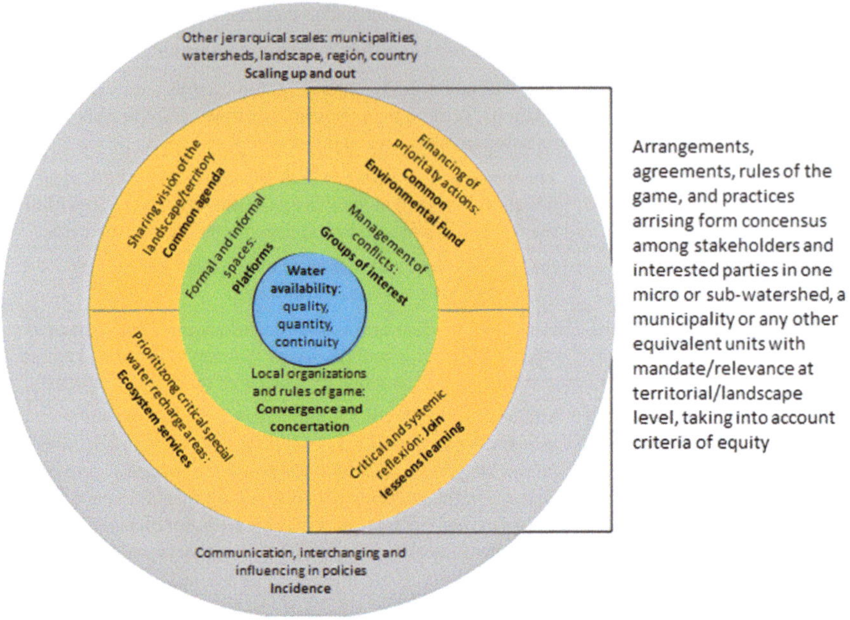

Fig. 10.1 Model of watershed's co-management promoted by CATIE. *Adapted from* Kammerbauer et al. (2010)

water recharge zones (Griebler and Avramov 2015; Kammerbauer et al. 2010; Leon y Prins 2010).

Considering that the model itself is a form of participation and management of local knowledge through action-research, it also incorporates critical and systematic reflection of the entire process, which leads to social learning. It also includes communication and exchange actions, addressing to policy influence to have an impact on the creation or enhancement of policies at the municipal and national levels (Jimenez y Benegas 2019). Finally, the model considers and integrates the institutionalization of watershed co-management through scaling up downward and upward at territorial scales, such as ecosystems, watershed, landscape, municipality, country, and region (Kammerbauer et al. 2010; Jimenez y Benegas 2019; Kauffman 2017).

10.4.1 Formulation of Watershed Management Plans According to CATIE's Experience

The basic methodology to build a watershed management plan combines reviewing secondary information with primary data to start with a characterization and diagnostic of the watershed under interest. Biophysics and socioeconomic aspects are the components of these chapters, where the spatial analysis of the key variables must be included (morphometric and climatic variables, land use and zoning, climate change analysis, among others). The diagnostic phase is complemented by participatory workshops where the watershed's stakeholders define the problems and solutions at the time that prioritize the sites in the watersheds where such problems-solutions are present. Currently, there are some modeling programs which can assist in this prioritization process, for example, Resource Investment Optimization System (RIOS) tool (Vogl et al. 2017) which has been mainly used in the formulation and design of water funds (Vogl et al. 2017a). Nevertheless, validation processes with the watershed's stakeholders are crucial not only to verify the results of the model, but also to identify territorial barriers, social conflicts, or any contextual situations which can potentially limit the implementation of projects of good watershed management practices. Furthermore, local stakeholders can in most cases help to identify possible financial sources for the plan considering their own capacities.

Based in the diagnostic, the next step is to translate the solutions into programs and projects organized by common objectives. The number of programs varies according to the sizes of the watershed, but it is between 3 and 5 generally. For instance, a watershed management plan can have four programs related to: (1) Capacity building processes, (2) agricultural development, (3) water resource management, and (4) conservation of ecosystem services. A key section of the plan is the distribution and location of these programs with their respective projects within the watershed according to the territorial zoning previously done following the characterization phase. The management plan is composed by a set of programs which derive in several projects organized by common objectives, as well as by management and monitoring subjects.

To continue describing a "Watershed management Plan," it will be used an example, the Watershed Management Plan of the Jesus Maria River (CATIE-FONAFIFO 2011), one of the most degraded catchments in Costa Rica (Prins et al. 2017). In this case, a set of four programs were proposed, each one designed to give answer to the problems and opportunities detected both by secondary information (documents, literature, datasets, etc.) but also by primary information using participatory methods, all of them in search to accomplishing the specific objectives. These programs were: (1) Ensure that local organizations and institutions working on issues related to the management of the Jesús María River watershed, acquire tools and instruments to strengthen their organizations and are able to administer resources to implement the watershed management plan, with the goal of benefiting the families that inhabit the watershed and contribute to the sustainability of natural resources; (2) improve the socioeconomic capacity of men and women in the watershed, based on profitable agricultural, livestock, and forestry production, which integrates ecological and organic production, food security, and the conservation of natural resources; (3) to rehabilitate the soils and vegetation cover in the degraded areas, as well as to conserve and increase the remaining permanent vegetation cover so that they serve as a base in the sustainable agricultural and forestry production, generating at the same time ecosystem services, all under an approach to "produce conserving and conserve producing," the latter known as a CATIE's slogan; and (4) improve the quality, quantity and availability of the water resource, which is needed, among others, for the development of all productive activities, services, and the environment, satisfying mainly the water demands of rural and urban families.

There is one key section in the watershed management plan that has to do with organization to implement and/or execute the plan. In this section, there is a proposal to the cost which suppose the ideal functioning of an organization or group of workers who must specifically devote themselves to the plan; this is called the "Coordinator Committee" or "Implementing Unit," such denomination could be similar or equivalent to others given in the legislation and/or normative of the country object of the plan. For this organization which will be implementing the plan, it is crucial to provide a basic budget for them to operate, including communication activities as well as a seed-funds for project management and implementation.

Finally, the last key component of the watershed management plan is the design and inclusion of a monitoring and evaluation system, to assure not only the follow up needed for any processes and activities implemented as part of the plan, but also to communicate and spread as clear as possible all information generated for decision-making, allowing the possibility to detect, preferably in real time, the problems and drawbacks during execution, making possible to adjust whatever is needed timely.

Watershed management plans usually have a temporal horizon between 10–20 years. For the Jesus Maria River Basin management plan, it was set a horizon of 15 years, but considering process in the short, medium, and long term, reflecting priority projects to be implemented as the starting point for the implementation of the whole plan. The monitoring and evaluation system is included in the plan with the goal to lead institutions of the watershed to use it as a key tool to improve continuously through an operative planning, also verifying the performance of management actions and guiding toward activities to accomplish the desired impact and sustainability in the watershed.

In the particular watershed management plan showed as example in this section, there were three guidelines proposed to organize their program and projects' implementing strategy: (1) Strategy of communication and scaling up and out with its corresponding financial and management of resources, (2) strategy to catalyze and conduct the plan through and how the executive team should be, and (3) operative strategy to communicate and disseminate the plan at the local, national, and international level when possible.

10.5 The Way Forward

Two important paths to upgrade and work on the evolution of the concepts and approaches developed with IWRM and IWM are presented in this section. One path is the integration of them with emerging approaches like nature-based solutions, ecosystem-based adaptation, resilience and socioecological systems, and one water. The second path is the integration and identification of evidence on the clear contribution of both IWRM and IWM to the sustainable development goals.

10.5.1 Integration, Re-Interpretation, and Upgrading with Emerging Approaches

IWRM and IWM are both established before a series of emerging concepts and approaches (i.e., nature-based solutions, ecosystem-based adaptation, resilience of socioecological systems, one water) which overall share the same goals of these two previously developed concepts. Although IWRM and IWM are still in force and its relevance is broadly recognized, it is crucial to understand and present the specific similarities and complementarities of all of them, and more important, it is key to share the experience of such, for example, good watershed management practices that has been implemented for longer time and now are renewed as part of or contributing to the bases for these emerging approaches. Such approaches are described in Table 10.2, including a definition, similarities with IWRM and IWM, and examples of synergies in practices proposed to implement each approach.

10.5.2 Contribution to the Sustainable Development Goals (SDG)

Findings from twenty-one watershed management plans (elaborated since 2005 to date) that were prepared or advised by CATIE in Central America and the Caribbean are summarized here as an example of the relevance and challenges to overcome when combining the contribution of watershed management planning to the SDG. Based on the components and general approach of each plan, the approach of each one was compared with the 17 SDGs. In this exercise, it was identified for each plan, which indicators or elements of the plan would be responding to each SDG, indicating which of the specific targets and indicators proposed by the United Nations (UN 2020) the plans would be contributing to.

It was found that all the plans analyzed address SDGs 1, 2, 6, 9, 11, 12, 13, and 15. That is, their components, work axes, programs, projects, and/or indicators show concrete actions to reduce poverty, hunger, ensure clean water and sanitation, promote industry, innovation and infrastructure, sustainable cities and communities, promote responsible production and consumption, climate action, and the conservation of life of terrestrial ecosystems. On the other hand, SDGs 17, 10, and 14 were addressed in only 10, 14, and 19% of the plans, respectively. Of the 21 plans analyzed, 675 indicators considered most relevant to contribute to the registration of SDG targets defined by the UN were identified. From these indicators, 69% are biophysical and 31% are socioeconomic. However, if analyzed by plan, it was found that only 42% (9 of the 21) have higher number of biophysical indicators than socioeconomic ones. The rest have either a balance or a greater number of socioeconomic indicators.

Of the 21 plans analyzed, 17 of them have been implemented to some extent, so that, based on interviews with key actors linked either to the development and/ or implementation of these plans, and who continue in these territories leading and/

Table 10.2 Complementarities, differences, and overlapping aspects of IWRM, IWM, and emerging related approaches

Emerging approach	Definition	Similarities and differences with IWRM and IWM	Examples of synergies in practices proposed to implement the approach
Nature-based solutions (NbS)	Promoting of nature as a means of providing solutions to societal challenges, while contributing to the sustainable management of the natural environment (Eggermont et al. 2015; Keestra et al. 2018; White et al. 2021)	NbS, IWRM, and IWM all promote natural resource management. However, IWRM focuses in one natural resource: water	Constructed wetland as a practice to face pollution due to wastewater is an NbS. Such practice could and has also been part of the portfolio of programs and projects identified in IWM and IWRM initiatives. NbS, IWRM, and IWM based the selection of solutions or practices in the analysis of the problems in the landscape of interest
Ecosystem-based adaptation (EbA)	Is the use of biodiversity and ecosystem services as part of a strategy to help people adapt to climate change (Donatti et al. 2020)	In most contexts EbA and IWM could be both focused in biodiversity and water ecosystem services when the land use capacity and main problems in the watershed are related, for example, to conservation of groundwater recharge areas	Implementation of agriculture practices (e.g., agroforestry and soil conservation) is an example of EbA measure (Donatti et al. 2020), but also, it has been promoted as a best watershed management practice prior to the appearance of EbA approach (Perez and Tschinkel 2003)

(continued)

Table 10.2 (continued)

Emerging approach	Definition	Similarities and differences with IWRM and IWM	Examples of synergies in practices proposed to implement the approach
Building resilience on socioecological systems (RSES)	Resilience in socioecological systems (RSES) is a multifaceted process turning on emergent behaviors that are an outcome of the network interactions that occur in the landscapes and communities of such SES (Alessa and Kliskey 2020)	RSES, IWRM, and IWM all seek enhancing decision-making processes, empowering interested parties and co-management of their landscapes, which nowadays is redesigned under the RSES approach	The Water Resilience Pilot Program (WRPP) of the State of Colorado, USA, is an agreement which was established for the direction of traditional disaster recovery funds, for the first time, toward watershed coalitions working to restore disaster-affected watersheds with explicit resilience goals in mind (Koebele et al. 2020). This is a typical example of integrated watershed management main goals and motivation to implement watershed management plans from the beginning of such concept
One water	The idea of "One Water" is not new but is increasingly implemented in water resources management due to climate change impacts on freshwater resources. Through this approach managers are retaining or holding the water until needed, instead of sending water and stormwater downstream or moving it immediately off the land (Dorworth et al. 2021)	The concept of "One Water" is intended to demonstrate structured thinking about how each component of the hydrologic cycle can show the degree of interconnection, improving assessments (Jiang and McBean 2021) and overall management of water resources. This is a "recycled" concept and foundation of IWRM and IWM	Nutrient recovery, utility rebranding, natural treatment systems, pay for performance goal sharing are examples of one water initiatives ongoing under the Clean Water Services Agency in the Tualatin River Watershed located in Washington County, Oregon, USA (Howe and Mukheibir 2015). These types of activities are considered also as good watershed management practices in IWM and IWRM programs and projects

or accompanying these processes, it was found that 94% of the plans mention a use oriented to guide the activities developed in the watershed, while 50% of the plans have also managed to guide the planning and management of the municipality, environmental unit, regional government, or other permanent institution in the watershed.

Looking ahead to the last nine years remaining to achieve the SDG targets, planning and management of watersheds and any landscape scale should prioritize the following steps:

- Analysis of country commitments to the SDGs and linkage to other multilateral commitments. The role of multilateral environmental agreements in achieving the 2030 Agenda and Sustainable Development Goals (SDGs) is clear (United Nations Framework Convention on Climate Change (UNFCCC) and its Paris agreement, Convention on Biological Diversity (CBD), Convention to Combat Desertification and Drought (UNCCD), Sendai Framework for Disaster Risk Reduction 2015–2030), among others, acquired by the country, thus it must be identified the synergies, complementarities, opportunities, availability, and management of resources of all kinds.
- Analysis of the territorial planning instruments in force and aligned with the main objectives and goals of watershed management. It is well known that the formulation of watershed management plans should be framed, for example, in the country's development plans, regional and municipal plans, plans of protected area, among others, to guarantee legitimacy in aspects such as territorial zoning, as well as to identify potential budgetary allocations foreseen in the country.
- Incorporation in the planning formulation team, experts in gender, equity, and inclusion issues, who can guide the design and implementation of the plans from the social and economic point of view of the SDGs.
- Adequacy of the working methodologies for the formulation of watershed management plans so that they are broadly participatory, generating conditions for the active involvement of vulnerable groups and, equally important, the involvement of local decision-makers (mayors, governors, institutional delegates, etc.). The aim should be to validate and legitimize the watersheds plans to facilitate their appropriation by key stakeholders.
- Extension of the methodologies used in the formulation of plans to take the first steps of implementation, through a process of dissemination and communication of the plan at all levels, accompanied by effective capacity building. The latter is based on the characterization and diagnosis, as an initial part of the formulation of the plan.
- Identification of positive leadership in the territory, strengthening, and accompaniment of these, in order to "deliver the watershed management plan formulated in a participatory manner and properly validated." This way there are more probabilities to ensure its implementation after the exit of projects or technical advisors who were responsible for conducting these processes.
- A vision to the future is required with timely adjustments in approaches and strategies, with effective and active participation of local, public, and private stakeholders, local, and national governments, as well as international cooperation. The responsibility for watershed management is shared and requires that economic and financial management for its achievement be a common agenda and integrate key issues such as payment for environmental services, valuation

of externalities, business vision, and, in general, self-generation of income in the watersheds to better build resilience and sustainability for these territories.

10.6 Final Remarks

Integrated watershed management (IWM) and integrated water resource management (IWRM) are complementary concepts with slight differences. Both concepts remain valid and relevant, being evolving for at least 40 and 20 years respectively.

Part of this constant evolution requires the incorporation of key emerging concepts like nature-based solutions, ecosystem-based adaptation, resilience of socioecological systems, and one water. In some cases, the whole concept of integrated watershed management is considered as a type of ecosystem-based adaptation or a nature-based solution. However, from the approaches of IWM and IWRM, most of the practices promoted as options to implement such emerging concepts have been used for at least 40 years, known as best watershed management practices or practices for conservation of soils and water. Thus, lessons learn during the longer implementing, practice and the reflection period accumulated by IWM and IWRM represent and advantage to improve and overcome old known barriers to scale up those emerging concepts, which, consequently, reinforce the relevance of the water-related approaches.

At the same time, emerging concepts and global commitments like the sustainable development goals should be the way forward in terms of integration, evolution, and updating of the approaches of IWM and IWRM. And the best way of visualizing such integration is for example, analyzing the contribution of key instruments as the watershed management plans to the compliance of the SDG, and also, building the actual and next generation of developing projects with the principles and lessons learn with IWM and IWRM to operationalize the emerging approaches (i.e., NbS and EbA), which are gaining recent interest and funding support, together with the SDG commitments, and under the umbrella of the different global conventions where water-related approaches directly contribute.

References

Alessa L, Kliskey A (2020) People, perceptions, and process: multisystemic resilience in social-ecological systems. In: Multisystemic resilience. Oxford University Press, pp 165–180

Ayers J, Forsyth T (2009) Community-based adaptation to climate change. Environ: Sci Policy for Sustain Developm 51(4):22–31

Benegas L, Faustino J (eds) (2008) Seminario internacional de cogestión de cuencas hidrográficas: experiencias y desafíos (14–16 octubre, Turrialba, Costa Rica). Memoria. Turrialba, Costa Rica, CATIE, p 157

Benegas Negri L, Watler W, Ríos N (2017) Priorización de áreas para implementar buenas prácticas de manejo en cuencas costeras: un enfoque de optimización hidrológica y económica en función de los servicios ecosistémicos. Turrialba, Costa Rica, CATIE/FUNDECOR, pp 61. (Serie técnica. Informe técnico, no. 411)

Cardwell HE, Cole RA, Cartwright LA, Martin LA (2006) Integrated water resources management: definitions and conceptual musings. J Contemporary Water Res Educat 135(1):8–18

Castellón N, Prins C (2009) El Comité Ejecutivo de la subcuenca del río Jucuapa, Nicaragua: avances, alcances y aprendizajes. Turrialba, Costa Rica, CATIE, pp 72. (Serie Técnica. Boletín Técnico no. 37)

CATIE-FONAFIFO (2011) Plan de manejo de la Cuenca del Río Jesús María. CATIE FONAFIFO, pp 158

CGIAR Research Program on Water, Land and Ecosystems (WLE) (2014) Ecosystem services and resilience framework. CGIAR Research Program on Water, Land and Ecosystems (WLE). Colombo, Sri Lanka, International Water Management Institute (IWMI), pp 46. https://doi.org/10.5337/2014.229

Cherlet J (2012) Tracing the emergence and deployment of the 'Integrated Water Resources Management' paradigm. In: Paper presented at the 12th EASA Biennial conference, Nanterre, 10–13 July 2012, Unpublished document

Cohen A, Davidson S (2011) The Watershed approach: challenges, antecedents, and the transition from technical tool to Governance unit. Water Alternat 4(1)

Cohen-Shacham E, Walters G, Janzen C, Maginnis S (eds) (2016) Nature-based Solutions to address global societal challenges. Gland, Switzerland, IUCN, pp 8 + 97

Conca K (2006) Governing watercontentious transnational politics and global institution building (No. 363.61 C6)

Darghouth S, Ward C, Gambarelli G, Styger E, Roux J (2008) Watershed management approaches, policies, and operations: lessons for scaling up. In: Water Sector Board Discussion Paper Series No. 11. The World Bank, Washington DC

Dixon J, Easter W (1991) Integrated watershed management: an approach to resource management-Chapter 1. In: Easter W, Dixon J, Hufschmidt M (eds) Watershed resources management: studies from Asia and the Pacific. Institute of Southeast Asian Studies, pp 3–15

Donatti CI, Harvey CA, Hole D et al (2020) Indicators to measure the climate change adaptation outcomes of ecosystem-based adaptation. Clim Change 158:413–433. https://doi.org/10.1007/s10584-019-02565-9

Dorworth L, Widhalm M, Salazar K (2021) Climate change and the one water approach to water resources management. Purdue University Extension-Sea Grant Ilinois, Indiana, pp 4

Dourojeanni A, Jouravlev A, Chávez G (2002) Gestión del agua a nivel de cuencas: teoría y práctica. Santiago, CL, CEPAL. Serie Recursos Naturales e Infraestructura, vol 47. pp 83

Eggermont H, Balian E, Azevedo JMN, Beumer V, Brodin T, Claudet J, Fady B, Grube M, Keune H, Lamarque P, Reuter K, Smith M, van Ham C, Weisser W, Le Roux X (2015) Nature-based solutions: new influence for environmental management and research in Europe. GAIA-Ecol Perspect Sci Soc 24(4):243–248

FAO (2006) The new generation of watershed management programmes and projects. FAO Forestry Paper 150. Food an Agriculture Organisation of the United Nations, Rome

FAO (2007) Why invest in watershed management? Food an Agriculture Organisation of the United Nations, Rome

Faustino J, Jimenez F (2005) Guía para elaborar planes de cogestión de cuencas. Programa Focuencas II: innovación, aprendizaje y comunicación para la cogestión adaptativa de cuencas. Grupo temático de manejo de cuencas. ASDI-CATIE, Turrialba, Costa Rica, pp 22

Faustino J, Jiménez F (2005) El modelo de cogestión de cuencas hidrográficas en América Central. Disponible en: www.portalcuencas.net

Forsyth T (2013) Community-based adaptation: a review of past and future challenges. Wiley Interdisciplinary Rev: Climate Change 4(5):439–446

Füssel HM (2007) Vulnerability: a generally applicable conceptual framework for climate change research. Glob Environ Chang 17(2):155–167

GEF (Global Environment Facility) (2016) Pacific R2R Programme. Geoscience Division, Pacific Community, pp 24

Griebler C, Avramov M (2015) Groundwater ecosystem services: a review. Freshwater Sci 34(1):355–367
Guiang E, Lucero S, Isorena T, Astrera L (2021) Mainstreaming a ridge to reef approach for sustainable development in the Pacific: a practitioners' guide. The Pacific Community (SPC), Suva, Fiji, pp 124
Guo M (2014) Effective watershed management: planning, implementation, and evaluation. Hydrol Current Res 5:e119
GWP (Global Water Partnership) (2000) Integrated water resources management. Global Water Partnership Technical Advisory Committee (TAC). GWP, pp 71
Howe C, Mukheibir P (2015) Pathways to one water a guide for institutional innovation. e Institute for Sustainable Futures, University of Technology, Sydney and ForEvaSolutions as an account of work sponsored by the water environment research foundation (WERF), pp 28
IPCC (2007) Climate change 2007: impacts, adaptation and vulnerability contribution of working group II to the fourth assessment report of the IPCC
Jiang A, McBean E (2021) Sponge city: using the "One Water" concept to improve understanding of flood management effectiveness. Water 13(5):583. https://doi.org/10.3390/w1305058
Jimenez F (2005) Gestión integral de cuencas hidrográficas: enfoques y estrategias actuales. Recursos, ciencia y decisión. CATIE. ISSN 1659-1224. Disponible en: http://orton.catie.ac.cr/repdoc/a0252e/a0252e.pdf
Jiménez-Otárola F, Benegas-Negri L, (2019) Experiencias y contribuciones del CATIE al manejo y gestión de cuencas hidrográficas en América tropical. Revista de Ciencias Ambientales (Trop J Environ Sci). (Enero-Junio, 2019) 53(1). Consultada el 13 de julio del 2020, EISSN: 2215-3896. Disponible en: https://www.revistas.una.ac.cr/index.php/ambientales/article/view/11379/14687
Kammerbauer H, León JA, Castellón N, Gómez S, González JM, Faustino J, Prins K (2009) Plataformas de concertación: una apuesta por la gobernabilidad local en cuencas hidrográficas. Tegucigalpa, Honduras, CATIE, pp 71. (Serie Técnica. Boletín Técnico no. 31)
Kammerbauer H, León J, Castellón N, Gómez S, Faustino J, Prins C (2010) Modelo de cogestión adaptativa de cuencas hidrográficas. Propuesta conceptual basada en la revisión crítica de las experiencias en Honduras y Nicaragua. Recursos Naturales y Ambiente 56–57:117–122
Kauffman C (2017) In: Grassroots global governance: local watershed management experiments and the evolution of sustainable development. Oxford University Press, pp 288
Keesstra S, Nunes J, Novara A, Finger D, Avelar D, Kalantari Z, Cerdà A (2018) The superior effect of nature-based solutions in land management for enhancing ecosystem services. Sci Total Environ 1(610–611):997–1009. https://doi.org/10.1016/j.scitotenv.2017.08.077. Epub 2017 Aug 20 PMID: 28838037
Koebele EA, Crow DA, Albright EA (2020) Building resilience during recovery: lessons from Colorado's watershed resilience pilot program. Environ Manage 66(1):1–15
Lara C, León J, Alemán L, Prado A (2007) Creando institucionalidad local en la cogestión para el manejo adaptativo de cuencas: la experiencia de la Mancorsaric a través de la mesa sectorial de ambiente y producción (Mesap). Turrialba, Costa Rica, CATIE, pp 60. (Serie Técnica. Informe Técnico no. 359)
León J, Prins C (2010) Gestión territorial para la protección colectiva del agua: demarcación participativa de la zona productora de agua, Carrizalón, Honduras. Turrialba, Costa Rica, CATIE, pp 48
León JA, Morales VF, Castellón N, Benegas Negri LA, Kammerbauer H (2012) Guía metodológica para la construcción de una agenda territorial común. Serie Técnica. Boletín Técnico (CATIE). Número 51. CATIE, Turrialba, Costa Rica, pp 52
Liu S, Costanza R, Farber S, Troy A (2010) Valuing ecosystem services: theory, practice, and the need for a transdisciplinary synthesis. Ann New York Acad Sci 1185(1):54–78
Ozyuvaci N, Ozhan S, Görcelioðlu E (1997) Integrated watershed management for sustainable development of renewable natural resources. In: Paper presented at the XI world forestry congress, Antalya, Turkey

Park SE, Howden SM, Crimp SJ, Gaydon DS, Attwood SJ, Kokic PN (2010) More than eco-efficiency is required to improve food security. Crop Sci 50:S–132

Perez C, Tschinkel H (2003) Improving watershed management in developing countries: a framework for prioritising sites and practices. In: Agricultural, research and extension network. Network paper No. 129, pp 20

Prins C, Vega D, Watler W, Mata E, Barboza C, Zúñiga P, Benegas L (2017) Acciones conjuntas y buenas prácticas para la implementación de actividades de restauración, manejo de cuencas y resiliencia de fincas y paisajes ante el cambio climático. Síntesis para decisores. PB 24. CATIE, Turrialba, Costa Rica. 6 p

Qi H, Altinakar M (2013) Integrated watershed management with multi-objective land-use optimizations under uncertainty. J Irrig Drain Eng 139(3):239–243

Rahaman MM, Varis O (2005) Integrated water resources management: evolution, prospects and future challenges. Sustain: Sci Pract Policy 1(1):15–21

Ramakrishna B (1997) Estrategia de Extensión para el Manejo Integrado de Cuencas Hidrográficas: Conceptos y Experiencias. San José, IICA/ GTZ, pp 326. Consultado el 12 de julio del 2020. Disponible en: http://orton.catie.ac.cr/repdoc/A9786e/A9786e.pdf

Reyes BA, Paiz CR, Lira MR, López ND, Gómez SN (2008) Creación y funcionamiento del Comité de Cuenca Bimunicipal Aguas Calientes: sistematización de experiencias. Turrialba, Costa Rica, CATIE, pp 36. (Serie Técnica. Informe Técnico no. 369)

Tallis H, Vogl A (2012) Standardized science for secure water. Conservation gateway-the nature conservancy. Available in: https://www.conservationgateway.org/News/Pages/standardized-science-secu.aspx

UN (United Nations) (2020) The sustainable development goal report 2020. United Nations Department of Economic and Social Affairs. UN, pp 66. Available in: https://doi.org/10.18356/214e6642-en

UN (United Nations)-Convention Framework on Climate Change (2016) Report of the conference of the parties on its twenty-first session, held in Paris from 30 November to 13 December 2015. FCCC/CP/2015/10/Add, pp 36. Available in: https://unfccc.int/sites/default/files/resource/docs/2015/cop21/eng/10a01.pdf

UNEP (2021) Progress on integrated water resources management. Tracking SDG 6 series: global indicator 6.5.1 updates and acceleration needs). UNEP, pp 111. Available in: https://wedocs.unep.org/bitstream/handle/20.500.11822/36690/PIWRS6.5.1.pdf

USAID-SWP (2016) Water security tool kit (series #1–6). Available at: https://www.globalwaters.org/swp

Viner D, Ekstrom M, Hulbert M, Warner NK, Wreford A, Zommers Z (2020) Understanding the dynamic nature of risk in climate change assessments—a new starting point for discussion. Atmospheric Sci Lett 21(4):e958

Vogl A, Tallis H, Douglass J, Sharp R, Veiga F, Benitez S, Leon J, Game E, Petry P, Guimeraes J, Lozano JS (2016) Resource investment optimization system (RIOS). In: Introduction and theoretical documentation project, TNC (ed.), Stanford University, pp 107

Vogl A, Goldstein JH, Daily GC, Vira B, Bremer L, McDonald R, Shemie D, Tellman E, Cassin J (2017a) Mainstreaming investments in watershed services to enhance water security: barriers and opportunities. Environ Sci Policy 75:19–27

Wang G, Mang S, Cai H, Liu S, Zhang Z, Wang L, Innes JL (2016) Integrated watershed management: evolution, development and emerging trends. J Forestry Res 27(5):967–994

Water Science School (2019) Watersheds and drainage basins. Unitated State Geologica Survey-USGS. Available in: https://www.usgs.gov/special-topics/water-science-school/science/watersheds-and-drainage-basins

White C, Collier MJ, Stout JC (2021) Using ecosystem services to measure the degree to which a solution is nature-based. Ecosyst Serv 50:101330

Chapter 11
From Drought to Sustainable Agricultural Production. Historical Water Resource Management Experiences in the Quíbor Valley (Venezuela)

Wilmer Barreto, Yelitza García-Orellana, and Duilio Torres-Rodríguez

Abstract Water scarcity is one of the limitations that restricts agricultural production, and it has been aggravated by climate change, which increases the need to optimize irrigation systems. Among the successful experiences on water management are those of the Iberian Peninsula and Israel. In Latin America, successful experiences are less promoted, the management of water resources in the Quíbor Valley, north-western Venezuela, constitutes a successful example of water resource management in agricultural development. Historically, water management in the region was carried out empirically, since the arrival of settlers from the Canary Islands in the last century. This led to problems of overexploitation of aquifers and salinization of soils. Over the years, based on experience and integration of research institutes, the optimization and improvement of irrigation techniques was achieved, which contribute to soil conservation and mitigating soil salinization. This chapter describes the origin of the irrigation techniques implemented by the settlers, the technical requirements for their adoption, advantages, disadvantages, and their impact on agricultural production in the region.

Keywords Aridity · Irrigation · Salinization · Drought

11.1 Introduction

When analyzing the edaphic and climatic conditions of the Quíbor Valley, the initial impression is that the arid conditions are characterized by low rainfall and high evaporation. The xerophytic vegetation typical of arid and semi-arid zones leads to a low content of organic matter in the soil (Torres et al. 2017). One might conclude that

W. Barreto (✉) · Y. García-Orellana · D. Torres-Rodríguez
Universidad Centroccidental Lisandro Alvarado, Barquisimeto, Venezuela
e-mail: wjbarret@gmail.com

G. A. Corzo Perez et al. (eds.), *Aqua Vitae*, The Latin American Studies Book Series,
https://doi.org/10.1007/978-3-031-57731-4_11

any agricultural exploitation in this area is unfavorable. However, in recent years, particularly since the 1970s, this region has become the most important horticultural production area in Venezuela (Pierre and Betancourt 2007).

Despite the water limitations in the region, the soil conditions are favorable. This is due to the absence of leaching processes, which results in the accumulation of a high content of exchangeable bases. Many of these bases are essential nutrients for plants. Additionally, the soils have a high cation exchange capacity due to their geological origin (Torres et al. 2017). This allows for the retention of most of the nutrients required for the metabolic processes of plants.

Fortunately, the farmers who settled in the area in the 1950s clearly understood the limitations and potential of the land. They emphasized that in order to fully exploit its potential, it was necessary to secure sources of water. However, they encountered another limitation as the area not only had low rainfall but also lacked reservoirs to supply water resources (Olivares et al. 2016).

The need for land to cultivate crops piqued the interest of the Canary Islanders in the Quiboró lands, despite the aforementioned challenges that required them to exert more effort to increase productivity. To accomplish this, it was necessary to apply a range of technical expertise in hydrography, climatology, soil management, and even hydraulics. They began by cultivating vegetables, primarily tomatoes (*Solanum lycopersicum*), and onions (*Allium cepa*), as well as sugarcane (*Saccharum officinarum*). Their production was so successful that by the early 1960s, Quíbor had become an iconic location for agricultural production in Venezuela (Pastrán 2018).

The scarcity of water is a limiting factor for production, so the islanders had to develop irrigation systems to avoid depending solely on rainfall. This was not a new concept for them, as they came from lands with similar climatic characteristics. They primarily developed surface irrigation systems, which may seem inefficient in terms of water usage. However, if we analyze them from an edaphic perspective, they are ideal because they allow for the formation of a water reservoir in the soil. This serves two purposes: first, it provides water to the crops, and second, it helps to maintain soil moisture. It serves two purposes: first, it reduces the concentration of salts in the root zone, and second, it facilitates water penetration into the soil profile, which is known for its low infiltration rate (Torres et al. 2017).

With the growth of technological advances, new irrigation alternatives have emerged and been adopted by producers. Many of these alternatives have a high technological level, as noted by Delgado et al. (2011). These advances were established at the end of the twentieth century with the establishment of agroindustrial emporiums, from which the experience of drip irrigation arose. Drip irrigation offers the advantages of water saving and mitigates the risks of salinity affectation, which is one of the most critical problems in the area. The application of flushing sheets is unfeasible due to water scarcity, as noted by Villafañe et al. (1999).

At the beginning of the twenty-first century, the development of irrigation systems in highly technified agricultural systems has been combined with practices aimed at improving water retention in the soil. These practices include the use of polymer-based conditioners (Henríquez et al. 2003), drip irrigation systems in farmhouses to reduce evapotranspiration (Bravo and Ocampo 2021), mulching in open field crops, incorporation of organic matter (source), and crop rotation (Jaurixje et al. 2013).

Despite the progress made, the area still heavily relies on rainfall for irrigation. According to Delgado et al. (2011), a significant proportion of farmers in the area have low technological proficiency. This, coupled with the high rurality index, makes them highly susceptible to extreme droughts, as reported by Olivares et al. (2017). Although small producers have received financial and technical support from the hydrological company and technological management agencies in the area, the vulnerability to droughts remains a concern.

This chapter will describe the main irrigation systems that have been adopted in the Quíbor Valley since the beginning of agricultural exploitation. It will cover the technical and physical requirements necessary for their implementation, as well as the advantages and disadvantages associated with each system. Additionally, the chapter will explore the impact of these systems on the increase of agricultural production in the area.

The study of irrigation's past, present, and future is a crucial element in addressing the challenges faced by farmers in the Quíbor Valley. These challenges include the dangers posed by global warming and climate change to tropical ecosystems. These threats could lead to the expansion of arid zones and cause desertification problems (Estrada and Silva 2017). To combat these issues, new production models must be adopted that prioritize sustainable development and ensure the preservation of water resources.

11.2 Water Availability in the Quíbor Valley

In the Quíbor Valley, the availability of water is perhaps the greatest limitation for the successful implementation of irrigation systems. This is due to the scarce precipitation, which averages less than 600 mm per year. Figure 11.1 illustrates that the Valley is predominantly affected by extremely severe, very severe, and severe droughts (87%). As a result, agricultural producers are compelled to rely on irrigation for crop production (Olivares et al. 2016).

At the start of agricultural production in the Quíbor Valley during the 1950s, irrigation systems were superficial and relied on runoff water stored in lagoons. However, as the agricultural industry expanded, the aquifer was increasingly utilized as a quaternary fluvio-lacustrine type filling, which has been intensively exploited. The first deep tubular well for aquifer exploitation was drilled in 1956. Currently, 50% of the water used for irrigation is sourced from the aquifer. In 1995, groundwater extraction reached 35 Mm^3, while natural recharge was estimated at approximately 23 Mm^3/year. There are currently 70 active pumping wells, and over the last 40 years, the static level of the aquifer has dropped by more than 100 m due to overexploitation, as reported by Jégat and Mora in 2015.

The supply of water for irrigation is expected to decrease significantly, with availability dropping from over 250 Mm^3 Ha-1 to below 100 Mm^3 by 2025 (see Fig. 11.2).

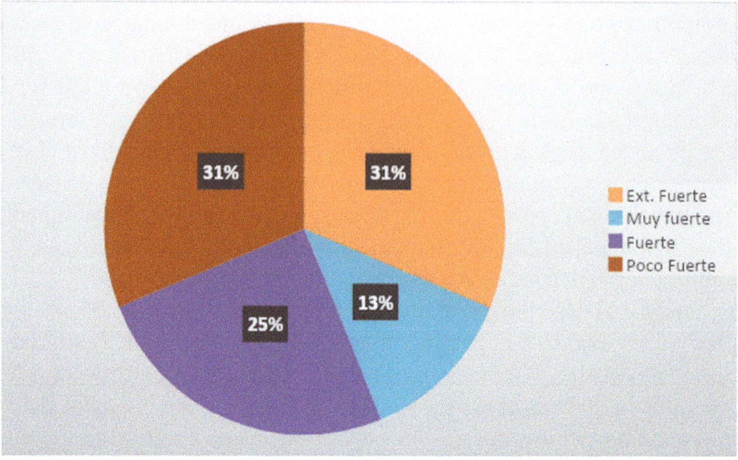

Fig. 11.1 Severity of drought in the Quíbor Valley. *Source* Olivares et al. (2016)

This depletion highlights the need to enhance irrigation systems and implement complementary practices to conserve water (Garduño and Nanni 2003).

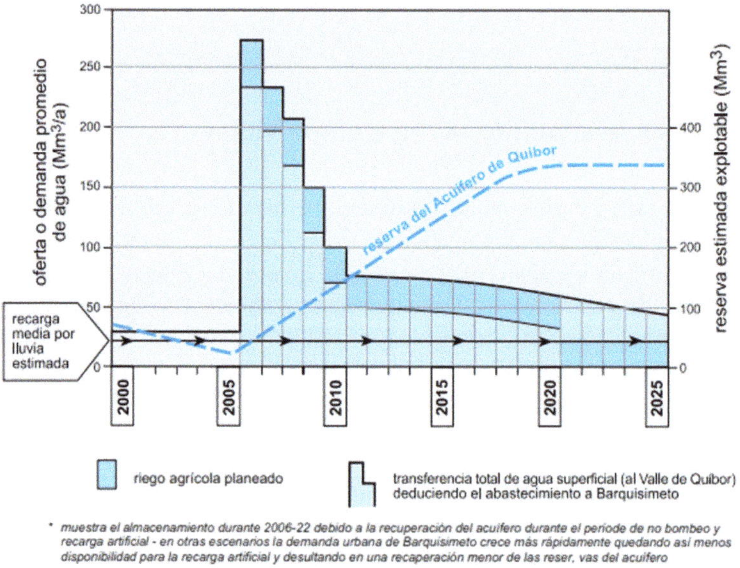

Fig. 11.2 Evolution of water reserves in aquifers of the Quíbor Valley. *Source* Garduño and Nanni (2003)

11.3 Water Quality and Soil Characteristics

One of the issues that can arise in the Quíbor Valley is poor water quality. Water extracted from wells often contains high levels of salt, leading to increased soil salinity and dispersion. These factors can negatively impact agricultural yields, causing a reduction in crop productivity. It is for this reason that most farmers combine water from wells with water from runoff that is stored in lagoons.

The main cause of aquifer depletion is that the rate of groundwater extraction for agricultural irrigation has exceeded recharge over a period of almost 40 years, currently estimated at about 27 Mm^3/year, which has caused a significant decrease in aquifer reserves with exploitation potential, from about 350 Mm^3 to 42 Mm^3, resulting in a continuous decrease in groundwater levels as well as an increase in groundwater salinity (Fig. 11.3). Both phenomena affect the productivity and profitability of local agriculture (Garduño and Nanni 2003).

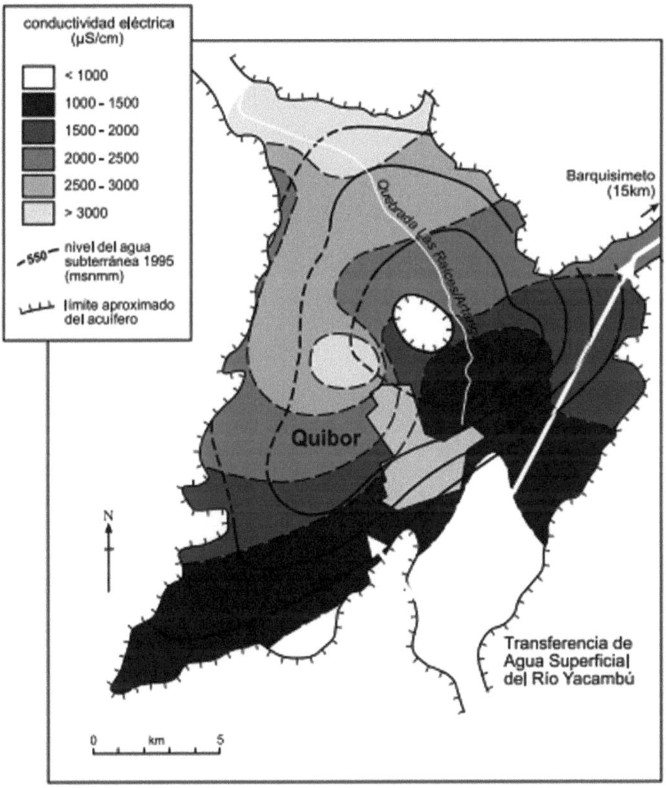

Fig. 11.3 Electrical conductivity of the Quíbor Valley aquifer waters (Coordinates not provided). *Source* Garduño and Nanni (2003)

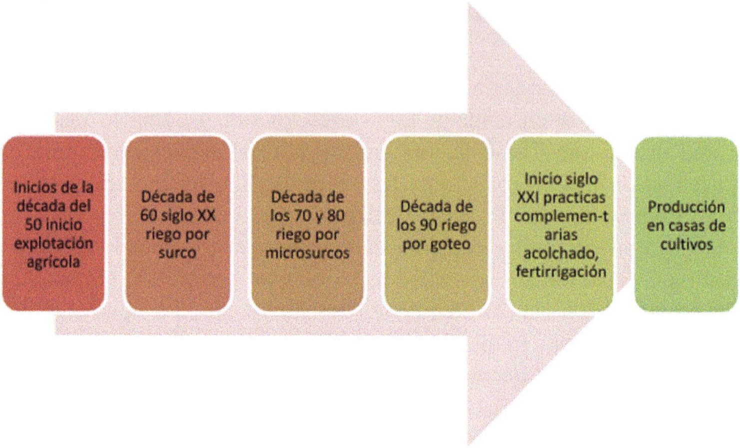

Fig. 11.4 Evolution of irrigation systems in Quíbor Valley

11.4 Irrigation in the Quíbor Valley

As mentioned above, the irrigation systems in the Quíbor Valley were initially based on surface irrigation methods, particularly furrow irrigation and its various variations. With technological progress, irrigation systems have advanced to localized methods such as drip irrigation. These methods are often accompanied by complementary practices aimed at reducing water evaporation, ultimately benefiting farmhouses (Fig. 11.4).

11.5 Coil Irrigation

Since the beginning of agricultural exploitation in the Quíbor Valley, farmers have realized that they are facing heavy soils with low infiltration, which is why a gravity irrigation system has emerged, which is a variant of the irrigation system called serpentine, whose main feature is that the water moves through short furrows in different directions (Fig. 11.5), and the length of the furrows varies from 2.50 to 5.00 m, with a width of about 0.60 m and a depth of 20 cm. There is no loss of water through runoff, and the water can remain on the furrow surface for up to four days (CIDIAT 1994).

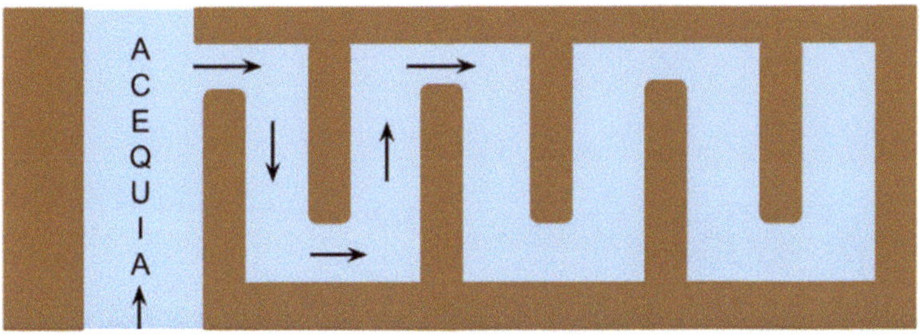

Fig. 11.5 Coil irrigation diagram. *Source* Garcia-Orellana (2002)

11.6 Irrigation by Drip Irrigation or Microsurcos

Due to the edaphic and climatic conditions of the Quíbor Valley, surface irrigation has been the predominant irrigation system in the area, with various modalities employed to optimize the use of water resources. This is due to the fact that water is the major limiting factor for agricultural production in the region. One of the irrigation methods that emerged in the 1980s was the brainchild of producer Mr. José Rodríguez. He modified furrow irrigation by reducing the distance between the furrows (as shown in Fig. 11.6), which in turn reduced the irrigation flow. This new method applied water to the soil through a "trickle" system. For this method to work, the furrows must have a slope between 1 and 2% (CIDIAT 1994).

This type of irrigation operates under the assumption that only the amount of water lost through daily evapotranspiration is replenished to the soil. In theory, this approach implies 100% efficiency. This type of irrigation is well-suited for heavy soils, but it requires good soil quality as it is highly sensitive to micro-topographic variations (Bohórquez et al. 1979).

It has been demonstrated that this method minimizes water loss due to evaporation by reducing the water surface area compared to the furrow irrigation system. The system has the great advantage of relatively low installation costs. Due to its ability to handle small water flows, the micro-furrow design has low maintenance costs (Moreno et al. 2014).

Another advantage of this irrigation system is that it requires minimal manpower. One person can irrigate several hectares simultaneously, unlike traditional methods where the irrigator must manually open and place "lids" to distribute water throughout the entire area.

Fig. 11.6 Microsurface irrigation in the Quíbor Valley. *Source* Own (2021)

11.7 Localized Irrigation

With the advancement of technology and the expansion of agricultural land, the Quíbor Valley faced a major challenge due to the scarcity of water resources. In the 1990s, to address the need for water conservation and combat the effects of salinization caused by overexploitation of the aquifer, localized irrigation techniques were developed. Although this irrigation system requires a high initial capital investment, it is a viable alternative for this area due to the unique characteristics of its soils, topography, and the scarcity and value of water (Rázuri et al. 2005).

Despite the efforts of several farmers to implement drip irrigation systems, it has not been widely accepted, with only 24.6% of farmers currently using it (Henríquez et al. 2010). There have been experiences with sprinkler irrigation in the area, particularly for forage crops, but there are no documented reports of these experiences.

The main reason for the lack of adoption of localized irrigation in the Quibor Valley is due to the predominant soil type in the area, which has a medium texture localized irrigation has not been adopted in the Quíbor Valley is due to the predominant soil type in the and is susceptible to compaction. Medium texture (FL) is susceptible to compaction. The superficial, lo que impide que of silt particles penetrate to form a goteo. In addition, the evaporation of water, the soil profile to the longer. In a drip irrigation which makes even though it cracks tends to remain in the soil during loses moisture. This can damage the time period. In addition, in an important part of the Quíbor Valley, the dominant type of clay is expansible, which causes the soil to crack as it loses moisture. This can cause damage to crop roots, loss of water in subsequent irrigations, and decreased yields.

According to a study conducted by Moreno et al. (2014) on caraota (*Phaseolus vulgaris L*) in the Quíbor Valley, the localized irrigation system proved to be more efficient in water application than the gravity irrigation system. However, when evaluating the yield of caraota, the gravity irrigation system produced better results. In the area, there are experiences of artisanal drip irrigation systems, which are an alternative for small producers who cannot have access to a conventional irrigation system, due to its high costs, in this context García-Orellana et al. (2012), proposed a low-pressure localized irrigation system that works with handmade drippers, which was designed by Olavarrieta (1997), as a solution to the irrigation problems of small producers in the semi-arid.

This system is characterized by using a water load of less than two meters, and the drippers are created by drilling holes in the hose or laterals smaller than 2 mm. This makes them extremely economical and leads to excellent results with the use of low-pressure localized irrigation. It has been demonstrated that handmade drippers provide adequate water supply for the cultivation of lechosa (*Carica papaya*) and parchita (*Passiflora edulis*) (García-Orellana et al. 2007, 2012). Its simplicity and low cost make it suitable for small-scale producers with limited economic and technical resources (Fig. 11.7).

Advantages of Irrigation Systems

The first irrigation system analyzed is known as the serpentine system, commonly used by farmers. This system minimizes water losses due to runoff, as the water remains on the furrow surface for an extended period, allowing enough time for infiltration to occur. Furthermore, it prevents soil cracking caused by expansible clays. Consequently, this irrigation system yields the highest crop yields, particularly for onions (*Allium cepa*).

The second irrigation method, also developed by local farmers, is the micro-furrow irrigation system. This method offers several advantages. Firstly, it minimizes water losses due to evaporation by reducing the water surface area compared to conventional furrow irrigation systems. Secondly, the installation costs are relatively low, as it

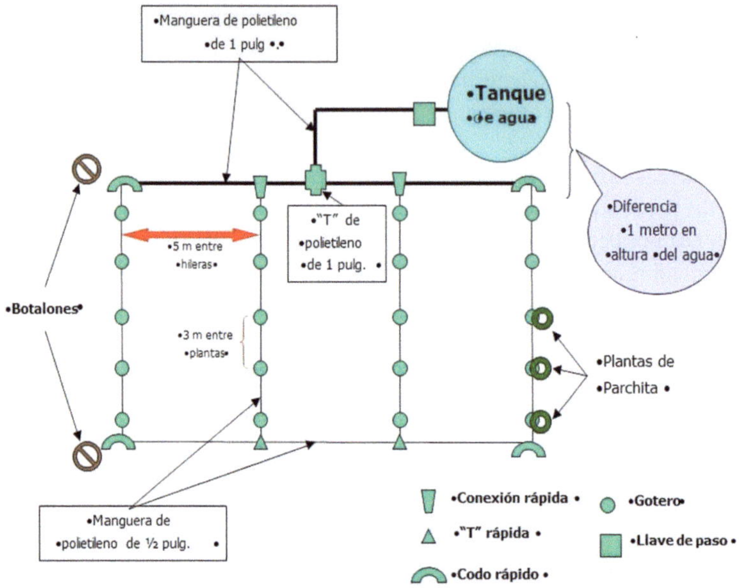

Fig. 11.7 Diagram of an artisanal irrigation system

only requires pumps, hoses, or pipes to transport water from the storage center to the plot being irrigated. Additionally, since the applied water flow is minimal, it does not damage the microsurface design, resulting in lower maintenance requirements compared to drip irrigation systems (Moreno et al. 2014).

The third system discussed is the drip irrigation system, which offers several advantages, including water conservation, control of salinity-related issues, and the ability to implement fertigation. Fertigation involves the application of water-soluble fertilizers, enhancing the efficiency of macro- and micro-nutrient usage. Moreover, this system reduces long-term fertilizer costs and mitigates soil salinization problems caused by excessive agrochemical use. It also enables the utilization of organic fertilizers derived from fungi, bacteria, and micro-algae.

Drip irrigation originated from traditional artisanal drip irrigation, providing the efficiency of the conventional system at significantly lower costs. This makes it a viable option for small and medium-scale producers in the Quíbor Valley (Olavarrieta 1997).

11.8 Disadvantages of Irrigation in the Quíbor Valley

The serpentine irrigation method has a few disadvantages. Firstly, it requires a significant amount of labor for both land preparation and irrigation operation. Additionally, a considerable amount of water evaporates from the furrow water surface, resulting in reported irrigation efficiencies of around 50% (CIDIAT 1994).

Regarding the micro-furrow irrigation system, despite its low flow rates, it requires a high frequency of irrigation. Given the limited availability of water in the Quíbor Valley, this becomes a disadvantage for this type of system.

The drip irrigation system, despite its numerous advantages, also has some disadvantages. Firstly, it has a high implementation cost and maintenance requirements. Moreover, managing drip irrigation, especially when combined with fertigation, requires a higher technological level. Consequently, it may not be feasible for most producers with limited technological resources.

On the other hand, the artisanal drip irrigation system is specifically designed for small-scale producers. However, due to its characteristics, it is not suitable for implementation in large areas and is not recommended for high-density crops such as onions.

11.9 Evaporation in Quíbor: Strategies for Reduction

Evaporation is one of the significant climatological factors that greatly impact agricultural production in the Quíbor Valley, with values exceeding 3000 mm per year. Consequently, the Quíbor Valley experiences distinct periods of drought, classified according to Olivares et al. (2016) as extreme drought (SPI-2 or less), severe intensity (SPI between -1.99 to -1.5), moderate (SPI between -1.49 to -1), and mild (> -1). As a result, farmers are compelled to transition from inefficient surface irrigation systems to localized irrigation systems to promote water conservation, particularly given the projected increase in frequency of extreme droughts due to climate change.

There have been experiences incorporating complementary practices such as mulching and the use of polymers to reduce evaporation. An example of such a system is illustrated in Fig. 11.8.

Another practice implemented in the Quíbor Valley to enhance soil quality and water movement in the soil involves the use of various substances. Authors such as Henríquez et al. (2000), Mogollón et al. (2015), and Torres et al. (2017) suggest that the application of organic acids, polysaccharides, synthetic, and organic conditioners can enhance the flocculation of clays. This, in turn, improves soil structural stability, enhances infiltration, and promotes water flow within the soil. In Quíbor, it has been observed that the issue is not only limited to water availability but also to its effective utilization by plants.

Fig. 11.8 Use of polymers in the Quíbor Valley. *Source* Anzalone (2008)

11.10 Physical Degradation and Soil Salinization: Management Strategies

The adoption of new irrigation systems in the Quíbor Valley has been driven by issues such as physical degradation and soil salinization, which are further exacerbated by the use of poor-quality water. One of the main challenges in the Quíbor Valley is the presence of sodium, which hinders water infiltration into the soil due to the damage it causes to the soil structure.

Additionally, the risk of salinization in Quíbor is high. Studies have found that 91.79% of samples from the area display values below 4 dS/m, while 8.21% exhibit higher values, indicating the presence of salts in the soil (Torres et al. 2017). Soil salinization is influenced by various technical factors, as depicted in Fig. 11.9, which are associated with the historical water management practices in the Quíbor Valley.

Irrigation and Agricultural Production in the Quíbor Valley

The necessity for irrigation in the Quíbor Valley arises from low and sporadic rainfall patterns, leading to a high reliance on irrigation water. This reliance is reflected in the irrigation/drought index. Over time, irrigation systems have evolved toward more efficient methods, resulting in increased productivity, particularly among technologically advanced producers.

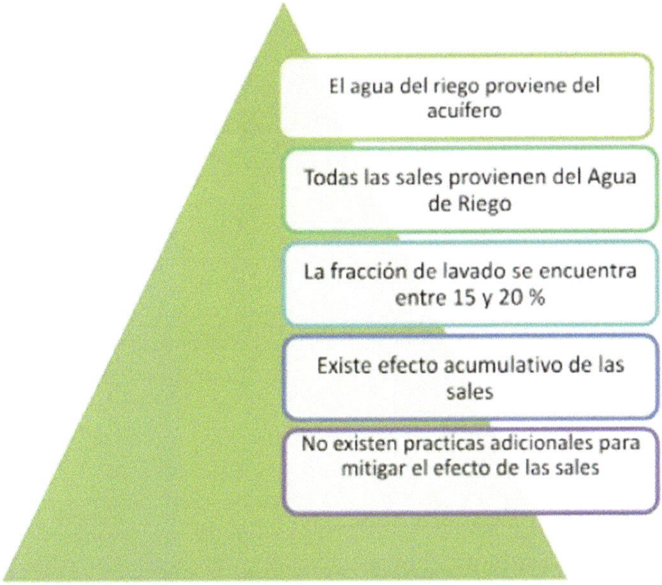

Fig. 11.9 Factors associated with historical water management leading to soil salinization

However, it is important to note that the Quíbor Valley also has a high rurality index, with many producers operating at a low level of technological advancement. This reality contributes to a moderate vulnerability index to drought (Table 11.1) since these producers may struggle to adopt more efficient irrigation systems or face limitations in implementing them. As a result, their yields are likely to be affected (Olivares et al. 2017).

Although gravity irrigation systems yield higher crop yields, from an economic perspective, the drip irrigation system is more beneficial for producers. This is because it utilizes a smaller volume of water, which is a limiting factor in production. The drip irrigation system demonstrates higher efficiency in water usage and offers additional advantages such as improved fertilizer utilization and reduced negative effects of salts on crops, particularly in sensitive crops like onions.

Table 11.2 compares the water use efficiency of both systems for the cultivation of caraota (*Phaseolus vulgaris*) under Quíbor Valley conditions. It clearly shows that while the gravity irrigation system achieves higher yields per unit area (1562,67 kg/ha), drip irrigation outperforms in terms of water use efficiency.

Table 11.1 Drought vulnerability and irrigation requirements for agricultural production in the Quíbor Valley

Irrigation/drought rate	Rurality index	Vulnerability index
0,83	0,24	0,51

Source Olivares et al. (2017)

Table 11.2 Comparison of yield and efficiency of irrigation systems in the cultivation of caraota (Phaseolus vulgaris L) under conditions of the Quíbor Valley

Irrigation system	Efficiency (kg ha^{-1})	Irrigation volume (m^3 ha^{-1})	Water use efficiency (kg m^{-3})
Microsurco	1562,67	7766,00	0,201
Dripping	1328,75	4449,57	0,270

Source Moreno et al. (2014)

Other studies have also evaluated the efficiency of irrigation systems on the vegetative growth of non-traditional crops in the Quíbor Valley (García et al. 2012). These studies found that the growth of passion fruit (*Passiflora edulis*) plants is comparable between drip irrigation and the traditional furrow system. However, the water applied in the localized irrigation system is only 54% of the volume used in gravity irrigation, indicating that localized irrigation provides better water economy.

Furthermore, García et al. (2007) previously discovered that water efficiency was significantly higher in papaya crops (*Carica papaya*) using handmade drip irrigation compared to the conventional drip irrigation system. The efficiency of the handmade system was found to be 1.24 kg/m^3, while the efficiency of the conventional drip irrigation system was 0.47. This resulted in higher yields in the handmade system, with a production of 7562.42 kg/ha compared to 2521 kg/ha in the conventional drip irrigation system. These findings demonstrate that this type of system can be a viable production alternative for small producers, especially for non-traditional crops in the Quíbor Valley.

Despite the promising results obtained in non-traditional crops, researchers are concerned about irrigation efficiency in large-scale crops like onions. López (2007) reports a water usage ranging between 447 and 457 mm for onion cultivation. While irrigation systems and complementary practices have advanced in terms of water use efficiency, research in the past decade has primarily focused on finding alternatives for managing soils affected by salts. The negative effects of salinity on yields are significant and are a consequence of the declining water quality due to the depletion of aquifer levels.

The development of irrigation systems encompasses a holistic approach that allows for the implementation of complementary practices. These practices aim to minimize not only water scarcity but also the structural deterioration caused by soil salinity. Such issues are commonly observed in the Quíbor Valley and have a noticeable impact on soil quality and agricultural yields. They reduce the availability of water by limiting its retention, infiltration, and flow within the soil profile.

Finally, to enhance water utilization, these systems have been complemented with the cultivation of crops such as paprika and onions in greenhouses. The use of greenhouses significantly reduces evapotranspiration, and the application of drip irrigation becomes more efficient by individually planting the crops in polyethylene bags. Additionally, the use of organic substrates promotes water retention and reduces evaporation.

11.11 Conclusions

In the Quíbor Valley, the evolution of irrigation systems has been a relatively recent process, starting in the mid-twentieth century. However, due to the characteristics of the farmers managing the farms, technological advancements have been rapid. These advancements have been primarily driven by the need to conserve water and cope with adverse climatic conditions that promote evapotranspiration processes.

In addition to the challenging climatic conditions, which have been further exacerbated by climate change, Quíbor farmers have had to adopt irrigation systems that not only ensure water savings but also help manage soils affected by salinity. Traditional practices such as applying leaching sheets to remove salts have become technically impractical. To address these challenges, irrigation practices in the Quíbor Valley have been accompanied by the use of physical soil amendments and crop rotation to mitigate physical degradation issues caused by high sodium levels.

Irrigation management in the Quíbor Valley has demonstrated a high level of technological advancement, leading to increased agricultural production. The area has become the most important horticultural zone in Venezuela, with notable yields of tomatoes, paprika, and onions. However, in some cases, there has been a gradual deterioration of soil quality, posing a threat to the productive potential of these lands.

Despite technological progress, the benefits of improved water conservation and usage have not reached small and medium-sized producers who continue to rely on rainfed agriculture, heavily dependent on the availability of rainfall. Their vulnerability has increased in recent years due to desertification, climate change, and the social and economic conditions affecting agricultural production in Venezuela.

References

Anzalone ÁL (2008) Evaluación de alternativas al uso del polietileno como cubierta del suelo para el manejo de malas hierbas y otros aspectos agronómicos en el cultivo del tomate (Lycopersicon esculentum P. Mill.) en España y Venezuela (Doctoral dissertation, Universidad de Zaragoza)

Bohórquez C, Rojo E, García G (1979) Riego por chorrito. Estación Experimental El Cují, CIARCO, Barquisimeto

Bravo FS, Ocampo GAR (2021) Crecimiento, evapotranspiración y uso de nutrientes en cultivo hidropónico de Eryngium foetidum, en dos diferentes ambientes y niveles de nutrición. Agronomía Costarricense 19–40. https://revistas.ucr.ac.cr/index.php/agrocost/article/view/45674

CIDIAT (1994) Evaluación de métodos de riego en el valle de Quíbor. Sistema Hidráulico Yacambú-Quíbor. Mérida, Venezuela, pp 149

Delgado A, Henríquez M, Guerra E, Torres D, Rodríguez V, Rodríguez O (2011) Tipología preliminar de los agricultores del Valle de Quíbor, Venezuela, según el uso de la tierra. Rev Fac Agron (LUZ) 28(Supl 1):688–698. https://www.revfacagronluz.org.ve/PDF/suplemento_diciembre_2011/

Estrada Gutiérrez G, Silva Hidalgo H (2019) Desertificación y Sequía. FINGUACH. Revista De Investigación Científica De La Facultad De Ingeniería De La Universidad Autónoma De

Chihuahua 4(12):12,13. Recuperado a partir de https://vocero.uach.mx/index.php/finguach/article/view/314

García-Orellana Y, Brito JJ, Gruber L (2007) Evaluación del efecto de dos variantes de riego localizado a baja presión en el cultivo de la lechosa en el Valle de Quíbor, Venezuela. Irriga 12(1):1–13. https://doi.org/10.15809/irriga.2007v12n1p1-13

García-Orellana Y, Brito JJ, Rivero J (2012) Evaluación del efecto de dos sistemas de riego sobre el desarrollo vegetativo del cultivo de la parchita maracuyá (Passiflora edulis var. flavicarpa Deneger) en el Valle de Quíbor, Venezuela. Irriga 17(2):264–273. https://doi.org/10.15809/irriga.2012v17n2p264

García-Orellana Y (2002) Determinación del requerimiento hídrico de la cebolla (Allium cepa L.) en condiciones edafoclimaticas del Valle de Quíbor, Edo. Lara. Trabajo de Maestría. Universidad de los Andes. Mérida, Venezuela

Garduño H, Nanni M (2003) Venezuela: Yacambú, Quíbor—un Proyecto para Integrar la Gestión del Agua Subterránea y el Agua Superficial. Caso 7. Banco Mundial. Programa asociado de la GW. 1–8. https://documents1.worldbank.org/curated/pt/891141468309541094/pdf/388070SPANISH0VE0GWMATE1CP10701PUBLIC1.pdf

Henríquez M, Montero F, Rodríguez O, Hernández A (2000) Effect of different concentrations of cactus dato, cactus lefaria, spanish tuna and PAM over some properties of a Quíbor-Lara soil. Revista De La Facultad De Agronomía, Universidad Del Zulia 17(4):295–306

Henríquez M, Rodríguez O, Montero F, Hernández A (2003) Efecto de acondicionadores naturales y sintéticos sobre los cationes solubles y la infiltración del agua en un Aridisol. Pesq Agrop Brasileira 38:311–316. https://doi.org/10.1590/S0100-204X2003000200019

Henríquez M, Rodríguez O, Rodríguez V, Guerra E (2010) Evaluación de las prácticas de descanso de los suelos que realizan los agricultores del Valle de Quíbor. Informe Técnico preliminar. SHYQ-UCLA, pp 20

Jaurixje M, Torres D, Mendoza B, Henríquez M, Contreras J (2013) Propiedades físicas y químicas del suelo y su relación con la actividad biológica bajo diferentes manejos en la zona de Quíbor, Estado Lara. Bioagro 25(1):47–56

Jégat H, Mora L (2015) Integrated land and water management in the Yacambu-Quíbor Project, Venezuela. In: Workshop on responding to the global food security challenge through coordinated land and water governance. Jointly organised by GWP, ILC and IWMI. Pretoria, South Africa, June, pp 15–16. https://www.gwp.org/globalassets/global/about-gwp/publications/proceedings-papers-pretoria-june-2015/6_herve_jegat_final.pdf

López J, Dennett M (2007) Estimación del uso del agua en el cultivo de cebolla (Allium cepa l.) En las condiciones de Quíbor, estado Lara, Venezuela. Bioagro 19(3):127–132. http://www.ucla.edu.ve/Bioagro/Rev19(3)

Mogollón JP, Vera MC, Martínez A (2015) Efecto de los plaguicidas sobre la calidad química y biológica del suelo en sistemas de producción de hortalizas del semiárido venezolano. Química Viva 14(1):69–89. https://www.redalyc.org/pdf/863/86340672008.pdf

Moreno MC, Silva G, Falcón C, Brito JJ, García-Orellana Y (2014) Evaluación del efecto de dos sistemas de riego en el cultivo de la caraota (Phaseolus vulgaris L) en condiciones del Valle de Quíbor, Estado Lara. IRRIGA 19(4):641–654. https://revistas.fca.unesp.br/index.php/irriga/article/view/688

Olavarrieta S, Artesanal R (1997) UCLA. Núcleo Universitario Juan A. de la Torre Material de Apoyo Docente., Carora, Venezuela, pp 15

Olivares BO, Cortez A, Rodríguez MF, Parra RM, Lobo D, Rey JC (2016) Análisis temporal de la sequía meteorológica en localidades semiáridas de Venezuela. UGCiencia 22:11–24. https://revistas.ugca.edu.co/index.php/ugciencia/article/view/481

Olivares B, Cortez A, Lobo D, Parra R, Rey J, Rodríguez M (2017) Evaluación de la vulnerabilidad agrícola a la sequía meteorológica en diferentes localidades de Venezuela. Revista De La Facultad De Agronomía De La Universidad Del Zulia 34(1):103–129. Recuperado a partir de https://produccioncientificaluz.org/index.php/agronomia/article/view/27222

Pastrán Calles FR (2018) Cultura e intercultural en Latinoamérica: presencia de los inmigrantes canarios en el estado Lara, 1961–1981. Red de Investigación Educativa 10(1):15–28. https://revistas.uclave.org/index.php/redine/article/view/1213/521

Pierre F, Betancourt P (2007) Residuos de plaguicidas organoclorados y organofosforados en el cultivo de cebolla en la depresión de Quíbor, Venezuela. Bioagro 19(2):69–78. http://ve.scielo.org/scielo.php?script=sci_arttext&pid=S1316-33612007000200002

Rázuri LR, Romero E, Galindo Rodríguez A, Hernández JD, Rosales JG, Linares de Gómez R (2005) Efecto de la densidad de siembra sobre el rendimiento de la cebolla, variedad americana, con riego por goteo, en el Valle de Quíbor. Agricultura andina 10(E):9–22. http://www.revencyt.ula.ve/storage/repo/ArchivoDocumento/agri/v10/art2.pdf

Torres D, Álvarez J, Contreras J, Henríquez M, Hernández W, Lorbes J, Mogollón JP (2017) Identificación de potencialidades y limitaciones de suelos agrícolas del estado Lara, Venezuela. Bioagro 29(3):207–218. https://www.redalyc.org/pdf/857/85752807006.pdf

Villafañe R, Abarca O, Azpúrua M, Ruiz T, Dugarte J (1999) Distribución espacial de la salinidad en los suelos de Quíbor y su relación con las limitaciones de drenaje y la calidad del agua. Bioagro 11(2):43–50. http://www.ucla.edu.ve/Bioagro/REV11(2)/1.%20Distribuci%C3%B3n%20espacial.pdf

MIX
Papier aus verantwortungsvollen Quellen
Paper from responsible sources
FSC® C105338

If you have any concerns about our products,
you can contact us on
ProductSafety@springernature.com

In case Publisher is established outside the EU,
the EU authorized representative is:
Springer Nature Customer Service Center GmbH
Europaplatz 3, 69115 Heidelberg, Germany

Printed by Libri Plureos GmbH
in Hamburg, Germany